Mathematical Analysis

Mathematical Analysis

Editors
Vishwambhar Pati • Alladi Sitaram

INDIAN ACADEMY OF SCIENCES

Universities Press

MATHEMATICAL ANALYSIS

UNIVERSITIES PRESS (INDIA) PRIVATE LIMITED

Registered Office
3-6-747/1/A and 3-6-754/1, Himayatnagar
Hyderabad 500 029, Telangana, India

Distributed by
Orient Blackswan Private Limited

Registered Office
3-6-752, Himayatnagar, Hyderabad 500 029, Telangana, India

Other Offices
Bengaluru, Bhopal, Chennai, Guwahati, Hyderabad, Jaipur, Kolkata, Lucknow, Mumbai, New Delhi, Noida, Patna, Vijayawada

First Published in 2001
Reprinted 2004, 2018

501506

ISBN: 978-81-7371-291-3

Typeset by
Devi Information Systems, Chennai

Printed in India at
Graphica Printers & Binders
Hyderabad- 5000 013

Published by
Universities Press (India) Private Limited
3-6-747/1/A and 3-6-754/1, Himayatnagar,
Hyderabad 500 029, Telangana, India.

Contents

Foreword

The Indian Academy of Sciences launched *Resonance* as a monthly journal devoted to science education in January 1996. *Resonance* is aimed largely at undergraduate students and teachers of science, though material of interest to somewhat younger students is also included. Each issue contains papers that span a wide area of science and mathematics, in various formats. Some are individual general articles, others consist of series with several parts. An effort is made to ensure a good expository quality in all of them.

"Echoes from Resonance" is a series of books born out of *Resonance*, by putting together in a coherent manner a collections of articles (both series and single pieces) taken from *Resonance*, all written around a common theme. Typically, the individual articles would have appeared quite independently at different times. These collections should prove useful to a reader who is keen to learn about a specific subject, with accounts given by different authors from different perspectives, but all in an expository manner. We hope these volumes would be useful for students and teachers alike, and that they will complement the structure of the individual issues of *Resonance*, which cover the different areas of science and mathematics in a balanced manner.

N. MUKUNDA

Introduction

The exposition of higher mathematics to young students always poses a challenge. A very formal style tends to obscure the underlying ideas, whereas an excessively informal style of presentation might slur over genuine technical difficulties. We feel that many of the mathematical articles that have appeared in *Resonance* so far have succeeded in striking the right balance between these two approaches. This is a selection of reprints from *Resonance*, published during the period 1996–98.[1] They are all in the area of mathematical analysis, and we have chosen those which broadly centre around the theme of sequences and series of numbers as well as functions.[2] It would, of course, be too ambitious to attempt a definition of mathematical analysis, but regardless of one's particular point of view, sequences and series play a very central role in the subject.

The contents are divided into three parts, with Part 2 constituting the core of the subject matter of this volume. In any mathematical endeavour, one has to come to grips with foundational issues, such as induction and the nature of infinity. These issues very often have to be addressed in the course of providing rigorous proofs of mathematical results, and hence we have chosen to include the articles that constitute Part 1. In Part 3, we have included short biographical sketches of some of the great mathematicians who have made major contributions to analysis. Needless to say, Part 3 makes no claims to completeness, and merely restricts itself to some of the great analysts who have made an appearance in the pages of *Resonance*. The article on Ramanujan is not, strictly speaking, a biographical sketch. It is a review of the best-selling book "The Man Who Knew Infinity", by R. Kanigel. In the course of his review, the author does provide interesting glimpses into the life of Ramanujan. The other great Indian mathematician of recent times is undoubtedly Harish-Chandra. His monumental contributions to harmonic analysis on Lie groups ought to be viewed as far-reaching generalisations of the ideas of Fourier series and transforms, which feature in Part 2 of this volume. Ironically, he remains largely unknown in the land of his birth. Therefore we thought it appropriate to include a sketch of his life.

Perhaps the first encounter that all of us have with infinite series and sequences is in the decimal expansion of a real number. For example, the realisation that the number $0.9999\ldots$(9 recurring) is the same as 1 requires the concept of the limit of a sequence, or equivalently, the sum of an infinite series—in this case, the geometric series. Historically, among the earliest uses of infinite series were the

---

[1] Please see page 139 for a list of *Resonance* issues where these articles were first published.

[2] The "Suggested Reading" in some of the articles in this collection may have some items from *Resonance*. So, first check whether such references are already in this collection. To facilitate your doing so, we give on page 139, details of the month, year, volume number, and page number of the *Resonance* issues in which the articles in this collection first appeared.

determination of the perimeter of a circle (and hence the value of π) by polygonal approximation, and the resolution of Zeno's famous paradox (a discussion of which may be found in many popular books on science, e.g., George Gamow's *One, Two, Three, Infinity*).

In the course of giving Euler's proof of the infinitude of primes, Shirali explains why the sum $\sum_{n=1}^{\infty} \frac{1}{n}$ is infinite, while the sum$\sum_{n=1}^{\infty} \frac{1}{n^2}$ is finite. The latter fact, as also the fact that every Cauchy sequence of real or complex numbers is convergent, is used in the article on Stirling's formula. (The reader is invited to discover for herself/himself exactly where!) It is worth noting that while Euler's proof of the infinitude of primes is much more complicated than the simple proof of Euclid's, it does give much more information about the distribution of primes, etc. Infinite sequences and series of numbers lead naturally to the notions of infinite sequences and series of functions, when one attempts to approximate an unfamiliar function f as a sum of familiar ones. Thus, one tries to find a natural sequence of functions $s_n(x)$ such that the given function $f(x)$ is the limit of the sequence of partial sums $\sum_{n=1}^{m} s_n(x), m \to \infty$. For example, in the familar Taylor expansion of $f(x)$ around a point a, $s_n(x)$ is just the polynomial $b_n(x-a)^n$, where $b_n = \frac{f^{(n)}(a)}{n!}$. Other ways of approximating $f(x)$ using Fourier series, and the new exciting technology of wavelets are dealt with in the articles of Thangavelu and Folland.

We have made this introduction as brief as possible, for as the Bard opines "Brevity is the soul of wit". So, read and enjoy!

Bangalore, August 2001 VISHWAMBHAR PATI AND ALLADI SITARAM

PART I

1

Mathematical Induction

An Impresario of the Infinite

B Sury

In the natural sciences, if a certain phenomenon is observed to occur a number of times, often a general law is formulated. This process is called *empirical induction*. In general, any reasoning that draws a general conclusion based on verification of particular cases is known as induction. But, in mathematics, a statement involving a natural number n might turn out to be erroneous even if it happens to be true for the first ten, or thousand, or even million natural numbers. For instance, the numbers $2^{2^0}+1 = 3, 2^{2^1}+1 = 5, 2^{2^2}+1 = 17, 2^{2^3}+1 = 257, 2^{2^4}+1 = 65537$ are all prime numbers and the 17th century mathematician Pierre de Fermat suggested that $2^{2^n} + 1$ must be prime for every positive integer n. However, a century later, another great mathematician Leonhard Euler showed that $2^{2^5} + 1 = 641 \times 6700417$. An even more convincing example is the following. If we evaluate the expression $991n^2 + 1$ for small values of n, the resulting number is not the square of a whole number. But, for $n = 12055735790331359447442538767$, the value is a perfect square. Indeed, this is the smallest value of n for which it is a square! This tells us that, in mathematics, a lot of care is needed to establish an induction procedure which proves a mathematical theorem for each of an infinite sequence of cases, without exception. The method of mathematical induction is such a procedure. Let us start with a simple example.

Suppose we want to prove the statement that $2^n > n$ for every natural number n. Clearly, this inequality holds for $n = 1$. Now, to prove the inequality for all natural numbers, we consider an *arbitrary* natural number $k \geq 1$. We *assume* that the inequality $2^k > k$ holds. Then, for the *next* natural number $k+1$, $2^{k+1} = 2\times 2^k > 2k$ by our *assumption* that $2^k > k$. Now, $2k = k+k \geq k+1$, so that the inequality $2^{k+1} > k + 1$ follows. Thus, we have proved that *if* the inequality is true for any particular k, *then* it is also true for $k + 1$.

The crux of the above argument rests on the points:

0) Given an infinite sequence of statements $P_r, P_{r+1}, \cdots$, we would like to prove that there is a 'next' to any statement, and each particular statement can be reached in a finite number of steps starting from the 'first' statement P_r.
1) There is a general method of proving that for any $n \geq r$, *if* P_n is true, *then* P_{n+1} is true; and
2) The first statement P_r is *known* to be true.

It is believed that these rules of logic are as fundamental to mathematics as the classical rules of Aristotelian logic.

It is necessary to verify both the steps (1) and (2) to avoid landing in absurdities. For example, if step (2) that 'starts induction' is not verified, one can 'prove' that all natural numbers are equal as follows. For, simply denote by P_n the statement '$n = n+1$'. Then, obviously, if P_n is assumed to be true, then $n = n+1$ and so $n+1 = n+2$, which means that P_{n+1} is also true.

Everybody has seen instances of mathematical induction being applied. The summing of arithmetic and geometric progressions are usually done by this method.

An important point is in order here. *Mathematical induction can be used to prove a statement that is given to begin with. As for coming up with that statement itself (as a guess, say), it is altogether a different matter. Therein lies the creative element which cannot be pinned down by any general rules.*

As we observed earlier, mathematical induction is a procedure that involves such extremely 'believable' logic that we accept it as valid reasoning. But, interestingly, we can actually prove its validity if we assume another believable principle which is that *any non-empty set of positive integers has a least number.* That this principle gives a proof of the validity of mathematical induction is left as an exercise to the reader.

We now proceed to give instances of various guises under which the method of mathematical induction appears and proves fruitful.

The following is a slight variant of the form in which induction is used.

To prove an infinite sequence $P_k, P_{k+1}, \cdots$, of assertions, one verifies the two steps:

i) P_k is true.
ii) For any $n \geq k$, if we assume that all the $P_k, P_{k+1}, \cdots, P_n$ hold good, then P_{n+1} also holds true.

Induction in Geometry

As an example, let us show that the sum of the interior angles of a (not necessarily convex) polygon of n sides is $180(n-2)$ degrees for all $n \geq 3$. Call this statement $P_n \cdot P_3$ is true as the sum for a triangle is 180 degrees. P_4 is also true since any quadrilateral can be split into two triangles.

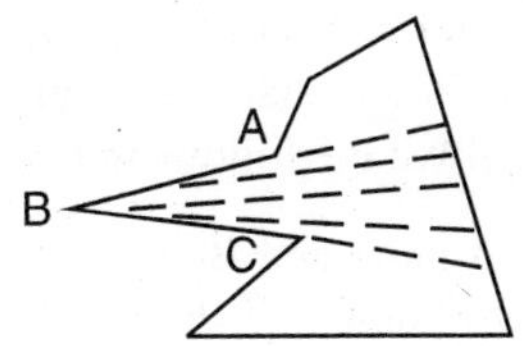

Figure 1.1

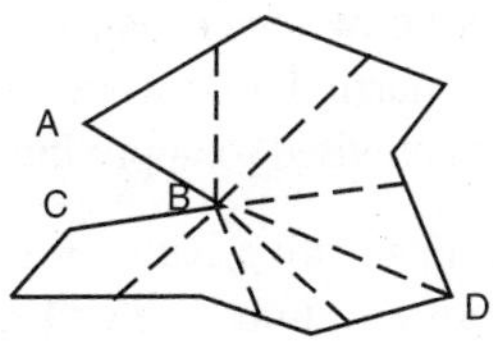

Figure 1.2

Now, let $n > 4$ and we assume that P_k is true for $k = 3, 4, \cdots, n-1$. Let $A_1, A_2, \cdots, A_n$ be the vertices of a polygon with n sides. We first notice that there is always a diagonal (i.e., a segment A_iA_j that is not a side) that splits the polygon into two with smaller number of sides. To see this, consider three neighbouring vertices A, B, C. Consider all the rays emanating from B and filling the interior angle ABC. We terminate any ray when it first meets a side or vertex of the polygon. Either all these rays intersect only one side (Figure 1.1) or they intersect more than one side (Figure 1.2). In the first case, AC is a diagonal that splits the original polygon into a triangle and a polygon with $n-1$ sides. In the second case at least one ray terminates on a vertex other than A or C. Call such a vertex D. Then, BD is a diagonal splitting the polygon into two with smaller number of sides.

Therefore, in general, let A_1A_k denote a diagonal which splits the polygon $A_1A_2\cdots A_n$ into the polygons $A_1A_2\cdots A_k$ and $A_kA_{k+1}\cdots A_nA_1$ of k and $n-k+2$ sides respectively. By the induction hypothesis, P_k and P_{n-k+2} are true, i.e., the sum of the interior angles of the original polygon $A_1A_2\cdots A_n$ is 180 $(k-2) + 180(n-k) = 180(n-2)$ degrees. So, P_n is true, which proves by induction that P_r is true for every $r \geq 3$.

After this standard example, we look at an example where it may not be quite apparent that induction can be used.

The Marriage Problem

The classical 'marriage problem' can be stated as follows. Suppose that each of a set of girls is acquainted with a subset from a given set of boys. Is it possible for each girl to marry one of her acquaintances?[1] Obviously, a necessary condition is that every set of m girls be collectively acquainted with at least m boys. *That this suffices is the assertion.* Here is a proof by induction.

Let n denote the number of girls. If $n = 1$, the assertion is trivial. If $n > 1$ and if it is true that every set of m girls, $1 \leq m < n$, has at least $m+1$ acquaintances, then an arbitrary girl is allowed her choice and the rest are referred to the induction hypothesis. If, on the other hand, some group of m girls, $1 \leq m < n$, has precisely m collective acquaintances, then this set of m girls is married off by induction and, it is indeed true that the rest of the $n-m$ girls satisfy the necessary condition with respect to the remaining boys. If this were not so, then some set of s spinsters with

[1] That we have interchanged the roles of girls and boys in our version is a miracle of modern genetics!

$1 \leq s \leq n - m$ know fewer than s bachelors, and this set of s spinsters together with the m just-married girls would have known fewer than $s + m$ boys.

The reader is invited to apply induction to solve the following two problems:

EXERCISE. (Consecutive Number Problem) Agatha and Beula are 'given' two consecutive natural numbers n and $n + 1$. Both know that the numbers are consecutive but neither knows whose number is bigger. After every minute a beep is heard and each is asked to simultaneously say out aloud whether she knows the other's number. Prove by induction on the smaller number n that the person who has the number n guesses correctly after precisely the n-th beep.

EXERCISE. (Macaulay Expansion) Given a natural number $d \geq 2$, let us write down the d-tuples of positive integers in a strictly decreasing order. Order the tuples lexicographically. Prove that the number of tuples appearing prior to a particular tuple $(k_d, k_{d-1}, \cdots, k_1)$ is precisely $\binom{k_d}{d} + \binom{k_{d-1}}{d-1} + \cdots \binom{k_1}{1}$.

This proves that any n has a unique expansion $n = \binom{k_d}{d} + \binom{k_{d-1}}{d-1} + \cdots \binom{k_1}{1}$ where $k_d > k_{d-1} > \cdots k_1$. Here $\binom{n}{r}$ denotes the binomial coefficient which is 0 when $n < r$.

Induction Incognito – Use of a 'Dummy' Element

Look at the following statement:

If $a_1 < a_2 < \ldots < a_{n+1}$ *are integers from the set* $\{1, 2, \ldots, 2n\}$, *then* a_i *divides* a_j *for some* $i < j$.

This can be solved by the 'pigeon-hole principle' as follows. Write $a_i = 2^{k_i} l_i$ with l_i odd. Then, $l_1, \ldots, l_{n+1}$ being $n+1$ odd numbers between 1 and $2n$, cannot be different. If $l_i = l_j = l$ with $i < j$, then, clearly $a_i = 2^{k_i} l$ divides $a_j = 2^{k_j} l$.

In terms of economy and elegance, this is unbeatable. However, we find to our surprise that even induction works and, in fact, proves the following more general statement:

S *Let* $r \geq 1$, *and let* $A \subset \{1, 2, \cdots, 2^r n\}$ *be a subset of cardinality* $(2^r - 1)n + 1$. *Then, there exists a chain of* $r + 1$ *elements of* A *with each dividing the next.*

Let us prove the original statement (corresponding to $r = 1$). Note that it is clearly true for $n = 1$. Assume it is true for n. Consider now $n + 2$ numbers $a_1 < \ldots < a_{n+2}$ among 1 to $2n + 2$. If $a_{n+1} \leq 2n$, we are done by the induction hypothesis. In the contrary case, we must have $a_{n+1} = 2n + 1$ and $a_{n+2} = 2n + 2$. If one of the a_i's is $n + 1$, we are done as it divides a_{n+2}. So, suppose $a_i \neq n + 1$ for each i. We may also assume that none of the n numbers $a_1, \ldots, a_n$ divides another, or else we have nothing to prove. Now, *we put in this new number* $n + 1$ (as a 'dummy element') to get $n + 1$ numbers between 1 and $2n$. By induction hypothesis, one of these $n + 1$ numbers divides another. Since this has happened only after the advent of the new number $n + 1$, it must be that either: (i) some $a_i (i \leq n)$ divides $n + 1$ or (ii) $n + 1$ divides some $a_i (i \leq n)$. But, clearly (ii) can not happen as $n+1 \neq a_n \leq 2n$. Thus, some $a_i (i \leq n)$ divides $n + 1$ and, therefore, divides $2n + 2 = a_{n+2}$ also. Thus, we used $n + 1$ as a 'dummy element' in this proof.

The reader is urged to complete the proof of the general statement along the same lines.

Now we come to a final example where induction appears in a different guise.

Backward Induction

If a statement is easily proved for a *particular* infinite subsequence of positive integers, it might be worthwhile to try whether 'backward induction' works. By this, we mean the following. Suppose we want to prove statements P_n for all positive integers n. Suppose, further, it is easy to check the veracity of P_n for all n in an infinite sequence of natural numbers. Then, if we check that for any $m \geq 2$, the truth of P_m implies the truth of P_{m-1}, the statements P_n follow for all positive integers n.

An instance is the familiar arithmetic mean–geometric mean inequality

$$P_n : \left(\sum_{i \leq n} a_i \right)^n \geq n^n \prod_{i \leq n} a_i$$

for arbitrary non-negative real numbers a_i, where equality holds if, and only if, all the numbers are equal.

On the one hand, we prove this for $n = 2^k$ by induction on k. Let $k = 1$. Then, $(a_1 + a_2)^2 \geq 4a_1 a_2$ with equality exactly when $a_1 = a_2$, since the difference $(a_1 + a_2)^2 - 4a_1 a_2 = (a_1 - a_2)^2$. Assume that P_n is true for $n = 2^r, r \leq k$. Let $a_i, i \leq 2^{k+1}$, be non-negative real numbers. Then, $\sum_{i \leq 2^{k+1}} a_i = \sum_{i \leq 2^k} b_i$ where $b_i = a_{2i-1} + a_{2i}$. Therefore,

$$\left(\sum_{i \leq 2^{k+1}} a_i \right)^{2^{k+1}} = \left(\sum_{i \leq 2^k} b_i \right)^{2^{k+1}} = \left(\left(\sum_{i \leq 2^k} b_i \right)^{2^k} \right)^2$$

$$\geq \left(\left(2^k\right)^{2^k} \prod_{i \leq 2^k} b_i \right)^2 = 2^{k2^{k+1}} \prod_{i \leq 2^k} b_i^2 \geq 2^{k2^{k+1}} \prod_{i \leq 2^k} (4a_{2i-1}a_{2i})$$

$$= 2^{k2^{k+1}} 4^{2^k} \prod_{i \leq 2^{k+1}} a_i = 2^{(k+1)2^{k+1}} \prod_{i \leq 2^{k+1}} a_i,$$

which proves that $P_{2^{k+1}}$ is true. Hence, by induction, P_{2^r} is valid for all $r \geq 1$. Moreover, note that the above proof also shows that the equality $\left(\sum_{i \leq 2^{k+1}} a_i\right)^{2^{k+1}} = 2^{(k+1)2^{k+1}} \prod_{i \leq 2^{k+1}} a_i$ implies that all the inequalities occurring on the way are equalities, which again proves by induction that equality can hold in P_{2^r} if, and only if, all the a_i's are equal.

On the other hand, for any m, the validity of P_m implies the validity of P_{m-1} as follows. Let $a_1, \cdots, a_{m-1}$ be given. Consider $a_m = \frac{1}{m-1}\sum_{i\leq m-1} a_i$. We may assume $a_m \neq 0$. Then $\sum_{i\leq m} a_i = \left(\frac{m}{m-1}\right)\sum_{i\leq m-1} a_i$ and

$$\left(\sum_{i\leq m-1} a_i\right)^{m-1} = \frac{1}{(m-1)a_m}\left(\sum_{i\leq m-1} a_i\right)^m$$
$$\geq \frac{1}{(m-1)a_m}\cdot\left(\frac{m-1}{m}\right)^m \cdot m^m \prod_{i=1}^m a_i$$
$$= (m-1)^{m-1}\left(\prod_{i=1}^{m-1} a_i\right).$$

Once again, by induction, equality implies that all the numbers are equal.

To end our discussion, the reader is invited to apply induction on the positive integer p below to prove the following result which solves an interesting two-player game called Euclid.

Let (p, q) be a pair of positive integers satisfying $p > q$. Each player subtracts a multiple of the smaller number from the bigger one without making the result negative. The winner is the one first hitting the highest common factor of p and q. Then there is a winning strategy for the first player if, and only if, $q < \frac{1}{2}(\sqrt{5}-1)p$.

Suggested Reading

R Courant and H Robbins. *What is Mathematics?* Oxford University Press, 1941.

L I Golovina and I M Yaglom. *Induction in Geometry*. Little Mathematics Library. Mir Publishers. Moscow, 1979.

B Sury
Statistics and Mathematics Unit
Indian Statistical Institute
Bangalore 560 059
Karnataka
Email: sury@isibang.ac.in

2

Nakula's Extraordinary Talent

Mathematics from the Mahabharatha!

C Musili

An attempt is made here to provide a mathematical model which can explain an intriguing talent of Nakula. Along the way, some simple and interesting properties of the real number system and related material are discussed.

Introduction

One of the exciting characters in the epic Mahabharatha, one that has fascinated me most since my childhood, is Nakula (the fourth of the Pandava brothers). Especially intriguing was his *ability to travel in rain riding on a horse, without getting wet.* Generally speaking, not everyone is aware of this great skill of Nakula as against the common knowledge that he could tame and train any kind of horse to any degree of perfection. None of the host of Sanskrit or Telugu pundits/scholars I asked, over the decades, could offer a plausible explanation. In fact, most of them looked surprised and wondered why I was seeking justification instead of leaving it aside as the product of a fertile imagination!

In this article, I would like to present a simple mathematical model as a possible motivation based on which Veda Vyasa might have attributed this extraordinary talent to Nakula. This will of course imply that Veda Vyasa was a mathematician! Surely. Why not? Even an Applied Topologist for that matter!

The Model

Assume that rain is falling on the floor sharply at points both of whose coordinates are rational numbers (with respect to a chosen set of coordinate axes). This means that it is raining rather *densely*. The simple theorem in topology about the *arcwise connectedness* of the complement of a *countable subset* of the Euclidean space $\mathbf{R}^n (n \geq 2)$ now implies that Nakula (shrunk to a point) can travel continuously from one dry point P to another dry point Q without getting wet. In fact, he can even travel smoothly, for example, along a suitable circular arc joining P and Q. There are *uncountably* many such circular arcs joining P and Q, implying that the choice is indeed *abundant*.

For those who can understand the words used above, there is no need to say more! Nevertheless, let me take the opportunity to explain the concepts used in the model for the benefit of others.

Real Numbers

> *The Real Number System is a set which can be geometrically thought of as the points on a straight line.*

We assume that the reader is familiar with the *'Real Number System'*. It is a set, denoted by $\mathbf{R}$, which can be geometrically thought of as the points on a straight line. For this reason, $\mathbf{R}$ is also referred to as the *'Real Line'*. It is a disjoint union of *'rational'* and *'irrational'* numbers. Rational numbers are the fractions m/n of integers m and n with $n > 0$. The set of all rational numbers is denoted by $\mathbf{Q}$. It is obvious that the sums and products of rational numbers are again rational. This is not true of the irrational numbers. The irrational numbers are not easy to describe or visualise. Some examples of irrational numbers (but not so obvious) are $\sqrt{2}, \sqrt{1.1}, \sqrt[3]{2}$ and so on.

Decimal Expansions

Perhaps, the simplest way to distinguish between the rational and irrational numbers is by means of the *decimal expansion* of real numbers, i.e., we assume that every real number x (rational or not) admits a decimal expansion in the form $x = x_0 \cdot x_1 x_2 \ldots x_m \ldots$, where x_0 is the *integral part* (i.e., the largest integer not greater than x, usually denoted by $[x]$) and $x_1, x_2, \ldots$ are the decimal digits of x all of which are integers lying between 0 and 9. This is easy to prove using simple properties of the convergence of series of non-negative real numbers. Now we have the following.

THEOREM. A real number $x = x_0 \cdot x_1 x_2 \ldots x_m \ldots$ is rational if and only if its decimal digits recur in a block of some size $n \geq 1$ beyond some stage m, i.e., the block of n digits $x_{m+1} x_{m+2} \cdots x_{m+n}$ recurs indefinitely after the mth decimal place. Indicating the recurring block under the bar, this is usually written as $x = x_0 \cdot x_1 x_2 \cdots x_m \overline{x_{m+1} x_{m+2} \cdots x_{m+n}}$.

In other words, a rational number is completely known in principle if its decimals are known for a sufficiently large number of places. The stage will of course depend on the particular number. For instance, $0 = 0.\overline{0}$, $1/3 = 0.\overline{3}$, $1/6 = 0.1\overline{6}$, $22/7 = 3.\overline{142857}$, $1/9 = 0.\overline{1}$, $1 = 1, \overline{0} = 0.\overline{9}$, etc., are all rational numbers, whereas numbers like $0.010110111011110\ldots, 0.1011001110001111 0000\ldots$, etc., are irrational.

What are the decimal expansions of $\sqrt{2}$, $\sqrt{11}$, $\sqrt[3]{2}$, π etc.? Despite the existence of the decimal expansions for real numbers, it may not be possible to give a rule for finding the decimal expansion of a given irrational number, be it as simple in appearance as $\sqrt{2}$, or as mysterious as π. Several mathematicians, including our legendary Srinivasa Ramanujan, wrote down the decimal expansion of π to a very large number of places.

Density of Rational Numbers

It is fairly easy to see that between any two distinct real numbers there are infinitely many rational as well as irrational numbers. In particular, given any real numbers x and a positive real number ε, there are infinitely many rational as well as irrational numbers between $x - \varepsilon$ and $x + \varepsilon$ (however small ε may be). This property means in mathematical terms that the subset of rational (or irrational) numbers is *dense* in the set of real numbers **R**.

Between any two distinct real numbers there are infinitely many rational as well as irrational numbers. This property means in mathematical terms that the subset of rational numbers is dense in the set of real numbers **R**.

Countable and Uncountable Sets

A set X is said to be *countable* if X is either finite or all of its elements can be *labelled* (or *enumerated*) as $x_1, x_2, \ldots, x_n, \ldots, \ldots$, labelled one for each of the natural numbers $1, 2, \ldots, n, \ldots, \ldots$. A set X is countable if its elements are counted in some order as we count the natural numbers. A set which is *not* countable is said to be *uncountable*. Uncountable sets are obviously infinite. Intuitively speaking, among infinite sets, countability or uncountability can be used as a tool to compare the infinitudes.

A set X is countable if its elements are counted in some order as we count the natural numbers.

Countability of Rational Numbers

The sets like the natural numbers $\mathbf{N} = \{1, 2, \ldots\}$, the integers $\mathbf{Z} = \{0, \pm 1, \pm 2, \ldots\}$, the rational numbers $\left\{\frac{m}{n} \mid m, n \in Z, n \neq 0\right\}$ etc., are all infinite of the same order, namely, countable. An enumeration for **Z** may be taken as for instance $\{0, 1, -1, 2, -2, 3, -3, \ldots, n, -n, \ldots\}$. An enumeration for **Q** can be taken as $\{0, x_1, -x_1, x_2, -x_2 \ldots, x_n, -x_n, \ldots\}$ if the subset of all positive rational numbers is enumerated as $\{x_1, x_2, \ldots, x_n, \ldots\}$. This is done diagrammatically as follows:

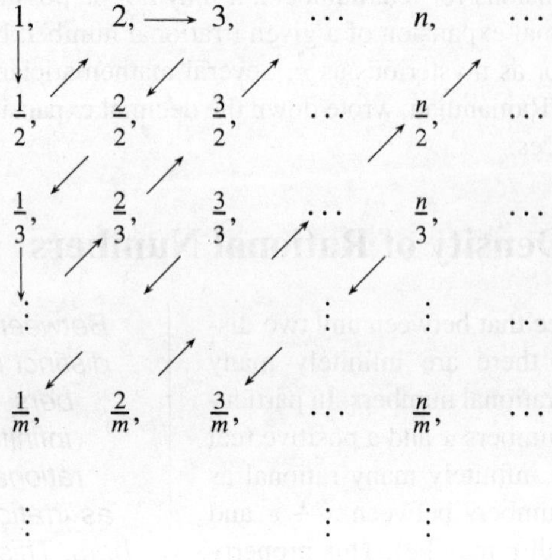

Finite Cartesian product of countable sets is countable but infinite Cartesian product of even finite sets need not be countable.

In this arrangement, every fraction $\frac{n}{m}$ appears at least once. Allowing repetitions (for example, 1 can appear again and again as 2/2, 3/3 and so on), we can enumerate them as

$$1; \frac{1}{2}, 2; 3, \frac{1}{3}; \frac{1}{4}, \frac{2}{3}, \frac{3}{2}, 4; \cdots, n-1; n, \frac{n-1}{2}, \frac{n-2}{3}, \cdots, \frac{2}{n-1}, \frac{1}{n}; \frac{1}{n+1}, \cdots$$

If $X = \{z_1, z_2, \ldots, z_n, \ldots\}$ is countable, so is the set $X \times X$ of pairs of elements (z_m, z_n) formed from the set X. An enumeration for $X \times X$ may be taken as

$$(z_1, z_1); (z_1, z_2), (z_2, z_1); (z_1, z_3), (z_2, z_2), (z_3, z_1); \ldots$$
$$\ldots; (z_1, z_n), (z_2, z_{n-1}), \ldots, (z_{n-1}, z_2)(z_n, z_1); \ldots$$

This arrangement is almost like what is done above for positive rational numbers. In particular, the set $\mathbf{Q} \times \mathbf{Q}$ of pairs of rational numbers is countable since $\mathbf{Q}$ is countable.

Some basic facts that can be proved on the same lines as above are the following:

1) A subset of a countable set is countable.
2) Countable union of countable sets is countable.

3) Finite Cartesian product of countable sets is countable but infinite Cartesian product of even finite sets need not be countable, as will be made clear below.

Uncountability of Real Numbers

Now for an example of an *uncountable* set, consider the set T of all real numbers between 0 and 0.2 consisting of those whose decimal expansion contains only the digits 0 and 1, i.e.,

$$T = \{0 \cdot a_1 a_2 a_3 \dots a_n \dots | a_i = 0 \text{ or } 1 \text{ for } i = 1, 2, \dots\}$$

For instance, T contains numbers like $0.\overline{0}, 0.\overline{10}, 0.10\overline{10}$, $0.010011000111000001111\dots, 0.\overline{1}$ etc.

What is obvious is that T is infinite but what is not at all clear is the fact that T is indeed uncountable! To see this, let T be enumerated if possible, as $T = \{d_1, d_2, d_3, \dots, d_n, \dots\}$. Let

$$\begin{aligned} d_1 &= 0 \cdot d_{11} d_{12} \cdots d_{1n} \cdots \\ d_2 &= 0 \cdot d_{21} d_{22} \cdots d_{2n} \cdots \\ &\vdots \\ d_m &= 0 \cdot d_{m1} d_{m2} \dots d_{mn} \cdots \\ &\vdots \end{aligned}$$

where each of the decimal digits $d_{11}, d_{12}, \dots; d_{21}, d_{22}, \dots$ is either 0 or 1. Now look at the number $x = 0 \cdot x_1 x_2 \dots x_r \dots$ where

$$x_1 = \begin{cases} 0 & \text{if } d_{11} = 1 \\ 1 & \text{if } d_{11} = 0 \end{cases} \dots x_2 = \begin{cases} 0 & \text{if } d_{22} = 1 \\ 1 & \text{if } d_{22} = 0 \end{cases} x_r = \begin{cases} 0 & \text{if } d_{rr} = 1 \\ 1 & \text{if } d_{rr} = 0 \end{cases} \cdots$$

This x is an element in the set T since its decimal digits are 0's or 1's. On the other hand, it differs from d_k at the kth decimal place (which is d_{kk}) for every k, which means that x has no place in the enumeration of the elements of T. This impossible situation arose because we assumed that T is countable.

The tiny-looking set T of the real numbers between 0 and $1/9 = 0.\overline{1}$ is uncountably infinite, and so the full set of all real numbers **R** which is much bigger than T must be uncountable. It follows that the set of all irrational numbers is uncountable since the rationals are countable. It can be seen easily that between any two distinct real numbers, there are uncountably many real numbers.

The set $\mathbf{R}^2 = \mathbf{R} \times \mathbf{R}$ of all pairs of real numbers (x, y) is uncountable because $\mathbf{R}^2$ is the Euclidean plane and it contains the uncountable set **R**, as a subset like the X-axis or the Y-axis or any straight line for that matter.

REMARKS. 1) The construction and proof of the uncountability of the set T above can be imitated to show that the set of all sequences formed from the two elements 0 and 1 is uncountable, i.e., the set

$$T = \{(x_n)_{n=1}^{\infty} | x_n = 0 \text{ or } 1 \text{ for } n = 1, 2, \ldots\}$$

is uncountable.

2) If $(A_n)_{n=1}^{\infty}$ is a sequence of non-empty finite sets, each having at least 2 elements, then the Cartesian product $\prod_{n=1}^{\infty} A_n$ is uncountable.
3) The *Cantor set*: A very well-known and popular example of an uncountable subset of the closed interval [0, 1] of the real line **R** is the so called *Cantor set* whose construction and properties are of independent interest.[1] It consists of all real numbers between 0 and 1 whose *trecimal expansion*, i.e., ternary expansion (i.e., expansion with base 3 instead of the decimal expansion with base 10) contains only the digits 0 and 2.

Density of Rational Points in the Plane

The subset $\mathbf{Q}^2 = \mathbf{Q} \times \mathbf{Q}$ of rational points is *dense* in the plane $\mathbf{R}^2 = \mathbf{R} \times \mathbf{R}$ in the sense that at each point $P = (x, y) \in \mathbf{R}^2$, and any circle C with center P and radius $\varepsilon > 0$, (however small ε may be), there are infinitely many rational points in the inside of C.

As before, this can be expressed as: *every neighbourhood of P in the plane contains infinitely many rational points, however small the neighbourhood may be.*

Pigeonhole Principle

Suppose there are m pigeons to be spread into n chambers. If $m > n$, then whichever way they are distributed, *at least one chamber* will contain 2 or more pigeons. This is called the pigeonhole principle for finite sets. Some natural extensions of this principle are the following:

1) Suppose there are infinitely many objects to be spread into finitely many boxes, then at least one box contains infinitely many objects. (Finite union of finite sets is finite.)
2) Suppose there are uncountably many objects to be spread into countably many boxes, then at least one box will contain uncountably many objects. (Countable union of countable sets is countable.)
3) Suppose there are countably many objects to be spread into uncountably many boxes, then only countably many boxes can be non-empty. (Uncountable union of disjoint non-empty sets is uncountable.)

---

[1] It is an *uncountable, closed, perfect, nowhere dense set of Lebesgue measure zero,* whatever these words mean!

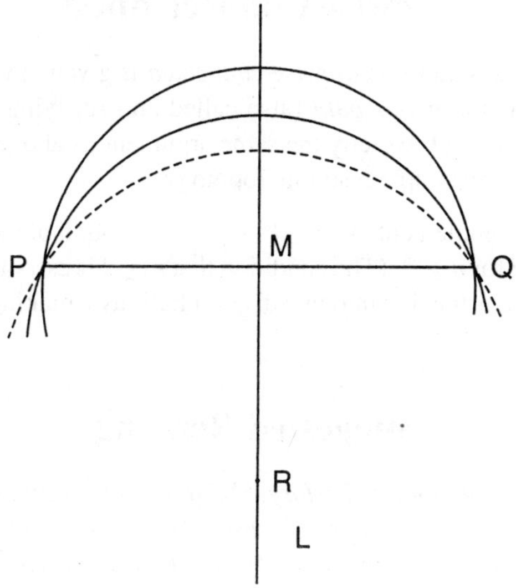

Figure 2.1

Justification of the Model

Assume that the rain is falling on the rational points $\mathbf{Q}^2$ in the plane $\mathbf{R}^2$. Take two dry points P and Q (i.e., $P, Q \in \mathbf{R}^2$ but $P, Q \notin \mathbf{Q}^2$) (Figure 2.1). Consider the set A of all circles in $\mathbf{R}^2$ passing through P and Q. This set A is uncountable, why? There is one and only one circle passing through P, Q and any point not on the line PQ. Or, if L is the perpendicular bisector of the segment PQ, there is one and only one circle passing through P and Q with each point R on L as the center and radius RP $(=RQ)$. The choice for R is uncountable and hence A is uncountable.

Since all these circles are different and pass through two non-rational common points P and Q, if a rational point S lies on one of them, it cannot lie on any other. In other words, a rational point lies on at the most one of these circles. Or equivalently, the set of circles passing through the rational points is countable.

Now apply the pigeonhole principle (3) above for the countable set of rational points $\mathbf{Q}^2$ and the uncountable set A of circles. Ignoring those countably many circles which pass through rational points (i.e., wet points), what is left in A is still uncountable. Nakula can choose any one of these circular arcs to go from P to Q.

(To be able to walk without stepping on or dashing against somebody in the streets of the mega cities like Calcutta or Mumbai, one has to be a Nakula indeed!)

Arcwise Connectedness

A subset X of $\mathbf{R}^n$ is said to be *arcwise connected* if given any two points P and Q in X, there is a *continuous path* (also called an *arc*) lying entirely in X and passing through P and Q. Exactly the same argument as above (replacing $\mathbf{Q}^2$ by Y) gives the following simple result in Topology.

THEOREM. The complement $X = \mathbf{R}^n - Y$ of a countable subset Y in R^n is *arcwise connected* for $n \geq 2$. (The condition that $n \geq 2$ is essential since removing a point from **R** breaks the line into two disjoint half rays, making it disconnected.)

Suggested Reading

O Ore. *Number Theory and its History*. McGraw–Hill. New York, 1948.
E Landau. *Foundations of Analysis*. Chelsea. NewYork, 1960.
M Klein. *Mathematical Thought from Ancient to Modern Times*. Oxford University Press. New York, 1972.
N H McCoy. *Introduction to Modern Algebra .(3rd Edition)*. Allyn and Bacon. Inc. Boston, 1975.
J R Munkres. *Topology — A First Course*. Prentice Hall of India. New Delhi, 1978.
M A Armstrong. *Basic Topology*. Undergraduate Texts in Mathematics. Springer–Verlag. New York, 1983.

C MUSILI
Department of Mathematics
and Statistics
University of Hyderabad
P.O. Central University
Hyderabad 500 046
Andhra Pradesh
Email:cmsm@uohyd.ernet.in
Fax:040-258120, 258145

3

Transfinite Numbers

What is Infinity?

S M Srivastava

In a series of revolutionary articles written during the last quarter of the nineteenth century, the great German mathematician Georg Cantor removed the age-old mistrust of infinity and created an exceptionally beautiful and useful theory of transfinite numbers. This is an introductory article on this topic.[1]

Introduction

How many points are there on a line? Which is more in number—points on a line or lines in a plane? These are some natural questions that have occurred to us sometime or the other. It is interesting to note the difference between the two questions. Do we have to know how many points and lines there are to answer the second question? Even primitive man could see whether he had more cows than his neighbour without knowing the actual number of cows they had—he would just pair off his cows with his neighbour's. So, to compare the size of two sets, we can start pairing off their elements. If one of them exhausts first we say that that particular set has fewer elements than the other one. Otherwise we say that the two sets have the same number of elements. If we are considering finite collections only, everything seems to be alright. Now imagine yourself walking into a hotel with infinitely many rooms which are all occupied. The receptionist shifts the guest in Room 1 to Room 2, the one in Room 2 to Room 3 and so on. Now Room 1 falls vacant and she gives it to you. As we can pair off the two sets of guests, the question arises: are there more guests in the hotel now? Thus

[1] The author would like to gratefully acknowledge the help received from S C Bagchi, T Krishnan, B V Rao and H Sarbadhikari while preparing this article.

we see that the idea of pairing off elements leads to counterintuitive results for infinite sets, e.g., an infinite set can have as many elements as a part of it. Georg Cantor (1845–1918) showed that this is a characteristic difference between finite and infinite sets and created an immensely useful branch of mathematics based on this idea which had a great impact on the whole of mathematics. For example, the question of what is a number (finite or infinite) is almost a philosophical one. However Cantor's work turned it into a precise mathematical concept.

Countable Sets

The simplest sets are those whose elements can be counted as the first, the second, the third etc. with the possibility that the counting may not stop. Such sets are called *countable*. Mathematically, a set A is countable if there is a one-one map from the set N of natural numbers $0, 1, 2, \ldots$, or from $\{0, 1, \cdots n-1\}$ (n being a natural number) onto A. (For $n = 0$ we take the set $\{0, 1, \cdots n-1\}$ to be the empty set.) If a set is not countable we call it *uncountable*. You can easily see that a set is countable if and only if its elements can be enumerated as $a_0, a_1, a_2, \ldots$, may be by repeating some of its elements. It follows that a subset of a countable set is countable.

EXAMPLES.

1) We can enumerate the set $N \times N$ of ordered pairs of natural numbers by the diagonal method as shown in the following diagram

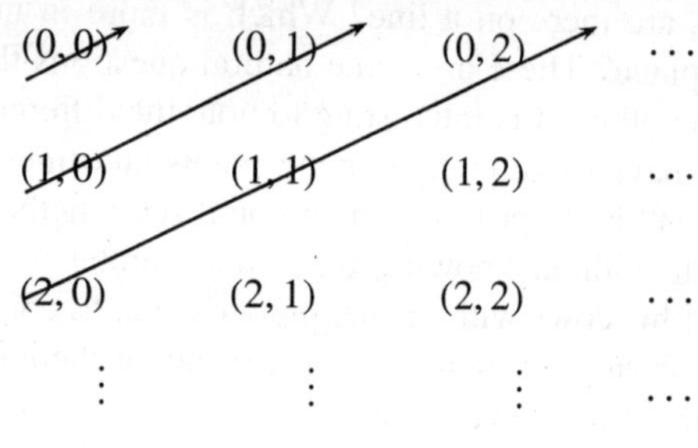

That is, we enumerate the elements of $N \times N$ as $(0, 0), (1, 0), (0, 1), (2, 0), (1, 1), (0, 2), \ldots$. By induction we see that the set of all k-tuples of natural numbers is also countable.

2) An argument similar to the one above shows that if $A_0, A_1, A_2, \ldots$ is a sequence of countable sets then their union $A = \bigcup_0^\infty A_n$ is also countable.

3) The set Q of all rational numbers is countable. (Exercise: Try to deduce this from the first example.)
The most natural question that arises now is: are there uncountable sets? The answer is yes. Here is an example.

4) If $a < b$ are real numbers, then the interval $[a, b]$ is uncountable.

PROOF. If possible, let $[a, b]$ be countable and let $\{a_0, a_1, \ldots\}$ be an enumeration of $[a, b]$. Define an increasing sequence $\{b_n\}$ and a decreasing sequence $\{c_n\}$ of real numbers in $[a, b]$ inductively as follows: Put $b_0 = a$ and $c_0 = b$. For some $n \in N$, let

$$b_0 < b_1 < \cdots < b_n < c_n < \cdots < c_1 < c_0$$

be defined. Let i_n be the first integer i such that $b_n < a_i < c_n$ and j_n be the first integer j such that $a_{i_n} < a_j < c_n$. Put $b_{n+1} = a_{i_n}$ and $c_{n+1} = a_{j_n}$. Now, the number $x = \sup \{b_n\}$ is in $[a, b]$ but is different from each of a_n. This is a contradiction.

Combining the above observations we easily see that every interval with more than one point contains uncountably many numbers which are not algebraic.[2] Such numbers are called *transcendental*. This simple argument of Cantor generated a lot of interest (as well as scepticism—for not giving an example of a transcendental number but only showing their existence) in set theory at that time. Later Hermite and Lindemann showed respectively that π and e are transcendental.

Equinumerosity—Order of Infinity

Say that sets A and B are *equinumerous* or are of the same *cardinality*, written $A \equiv B$, if there exists a one-one map f from A onto B.

EXERCISE. Show that if I and J are two intervals having more than one point (which may be bounded or unbounded, may or may not include their end points) they have the same cardinality.

In the last section we saw that the idea of equinumerosity, besides being a natural one, is useful too. So let us study it.

Say that sets A and B are equinumerous or are of the same cardinality, written A ≡ B, if there exists a one–one map f from A onto B.

Let X and Y be sets. The collection of all subsets of a set X is itself a set called the *power set* of X denoted by $\mathcal{P}(X)$. Similarly the collection of all functions from Y to X forms a set which we denote by X^Y. If A is a subset of X, then its *characteristic function* is the map $\chi_A : X \to \{0, 1\}$ where

$$\chi_A(x) = \begin{cases} 1 & \text{if } x \in A \\ 0 & \text{otherwise}, \end{cases}$$

You can easily verify that $A \to \chi_A$ defines a one-one map from $\mathcal{P}(X)$ onto $\{0, 1\}^X$.

So far, we have seen only two different 'orders of infinity': those equinumerous either with N, or with the set of real numbers $\mathfrak{R}$. Are there any more? Cantor's answer was simple but breath-taking.

[2] An *algebraic number* is a complex number which is a root of a polynomial having rational numbers for coefficients. For e.g.: $\sqrt{2}$ is an algebraic number; it satisfies $X^2 - 2 = 0$

Is there an uncountable set of real numbers of cardinality less than that of $\mathfrak{R}$?

Say that *the cardinality of a set A is less than or equal to the cardinality of set B*, written $A \leq_c B$, if there is a one-one function f from A to B. If $A \leq_c B$ but $A \not\equiv B$ then we say that *the cardinality of A is less than the cardinality of B* and symbolically write $A <_c B$.

Notice that $N <_c \mathfrak{R}$.

THEOREM. (Cantor) For any set X, $X <_c \mathcal{P}(X)$.

(Note that the power set of the empty set is not empty—it contains the empty set.)

PROOF. The map $x \to \{x\}$ from X to $\mathcal{P}(X)$ is one-one. Therefore, $X \leq_c \mathcal{P}(X)$. Now take any map $f : X \to \mathcal{P}(X)$. We show that f cannot be mapped onto $\mathcal{P}(X)$. To show this, consider the set

$$A = \{x \in X | x \notin f(x)\}.$$

(A may not have any point, thus it may be an empty set—yet a subset of X).

If $A = f(x_0)$ for some $x_0 \in X$, then note that

$$x_0 \in A \Leftrightarrow x_0 \notin A.$$

This contradiction proves our claim.

Now we see that

$$N <_c \mathcal{P}(N) <_c \mathcal{P}(\mathcal{P}(N)) <_c \cdots$$

Let T be the union of all the sets $N, \mathcal{P}(N), \mathcal{P}(\mathcal{P}(N)), \cdots$. Then T is of cardinality larger than each of the sets described above. We can now similarly proceed with T and get a never ending class of sets of higher cardinalities! Cantor at this point raises a very interesting problem. *Is there an infinite set whose cardinality is different from the cardinalities of each of the sets so obtained?* In particular, is there an uncountable set of real numbers of cardinality less than that of $\mathfrak{R}$? These turned out to be among the most fundamental problems not only in set theory but in the whole of mathematics. Cantor could neither solve these problems nor did he live long enough to see their surprising solution. We shall discuss these later.

The following result is very useful in proving equinumerosity of two sets.

SCHRÖDER–BERNSTEIN THEOREM. For any two sets X and Y,

$$X \leq_c Y \;\&\; Y \leq_c X \Rightarrow X \equiv Y.$$

PROOF. (Dedekind) Fix one-one maps $f : X \to Y$ and $g : Y \to X$. Consider the map $\mathcal{H} : \mathcal{P}(X) \to \mathcal{P}(X)$ defined by

$$\mathcal{H}(A) = X \setminus g(Y \setminus f(A)), A \subset X.$$

Then

i) $A \subset B \subset X \Rightarrow \mathcal{H}(A) \subset \mathcal{H}(B)$, and

ii) $\mathcal{H}\left(\bigcup_n A_n\right) = \bigcup_n \mathcal{H}(A_n)$

Now, let

$A_0 = \emptyset$, and

$A_{n+1} = \mathcal{H}(A_n), n = 0, 1, 2, \ldots$.

Let $E = \bigcup_n A_n$. Then, $\mathcal{H}(E) = E$. Define $h : X \to Y$ by

$$h(x) = \begin{cases} f(x) & \text{if } x \in E \\ g^{-1}(x) & \text{otherwise}. \end{cases}$$

The map $h : X \to Y$ is easily seen to be one-one and onto.

Here are some applications of Schröder–Bernstein theorem.

1) The function $A \to \sum_{n \in A} \frac{2}{3^{n+1}}$ from $\mathcal{P}(N)$ to $\mathfrak{R}$ is one-one. Therefore, $\mathcal{P}(N) \leq_c \mathfrak{R}$. On the other hand the map $x \to \{r \in Q | r < x\}$ from $\mathfrak{R}$ to $\mathcal{P}(Q)$ is one-one and so, $\mathfrak{R} \leq_c \mathcal{P}(Q)$. But $Q \equiv N$. Therefore, $\mathfrak{R} \leq_c \mathcal{P}(N)$. By Schröder–Bernstein theorem, $\mathfrak{R} \equiv \mathcal{P}(N) (\equiv \{0, 1\}^N)$.
2) Fix a one-one map $x \to (x_0, x_1, x_2, \ldots)$ from $\mathfrak{R}$ onto $\{0, 1\}^N$, the set of sequences of 0's and 1's. Then the function $(x, y) \to (x_0, y_0, x_1, y_1, \ldots)$ of the pairs of real variables to $\{0, 1\}^N$ is one-one and onto. So, $\mathfrak{R}^2 \equiv \mathfrak{R}$. Thus there are precisely as many points on the X-axis as in the entire plane! By induction on positive integers k you can now show that $\mathfrak{R}^k$ and $\mathfrak{R}$ are equinumerous. Similarly, using $N \times N \equiv N$, you can show that $\mathfrak{R}$ and $\mathfrak{R}^N$, the set of sequences of real numbers are equinumerous.
3) Show that the sets of points on a line and lines in a plane are equinumerous.

The Axiom of Choice

Are the sizes of any two sets necessarily comparable? That is, if X, Y are sets, is it true that at least one of the relations $X \leq_c Y$ or $Y \leq_c X$ holds? To answer this question, we need a hypothesis on sets, known as the axiom of choice.

THE AXIOM OF CHOICE (AC). If $\{A_i\}_{i \in I}$ is a family of non-empty sets, then there is a function $f : I \to \bigcup_i A_i$ such that $f(i) \in A_i$ for every $i \in I$.

Such a function f is called a choice function. Note that if I is finite, then by induction on the number of elements in I we can show that a choice function exists. If I is infinite, then we do not know how to prove the existence of such a map. The problem can be explained by the following example of Russell. Let $A_0, A_1, A_2, \ldots$ be a sequence of pairs of shoes. Let $f(n)$ be the left shoe in the n-th pair A_n and so the choice function in this case certainly exists. Instead, let $A_0, A_1, A_2, \ldots$ be a sequence of pairs of socks. Now we are unable to give a rule to define a choice function for the sequence of socks $A_0, A_1, A_2, \cdots$! AC asserts the existence of a function without giving any rule or any construction for defining such a function. Because of its non-constructive nature, AC met with serious criticism. For an excellent account of AC see Gregory Moore's book mentioned in Suggested Reading. We only remark that AC is indispensable not only for the theory of cardinal numbers but for the whole of mathematics. Note that while proving that the union of a sequence of countable sets $A_0, A_1, \ldots$ is countable we have used AC. For each n, we chose an enumeration of A_n. But there are infinitely many such enumerations and we did not specify any rule to choose them. The next result shows that every infinite set has a proper equinumerous subset. In its proof we use AC.

THEOREM. If X is infinite and $A \subset X$ finite then $X \backslash A$ and X have the same cardinality.

PROOF. By AC, fix a choice function $f : \mathcal{P}(X) \backslash \{\emptyset\} \to X$ such that $f(E) \in E$ for every non-empty subset E of X. Let $A = \{a_0, a_1, \ldots, a_n\}$. Inductively define $a_{n+1}, a_{n+2}, \ldots$ such that

$$a_{n+k+1} = f(X \backslash \{a_0, a_1, \ldots, a_{n+k}\}).$$

Now define $h : X \to X \backslash A$ by

$$h(x) = \begin{cases} a_{n+k+1} & \text{if } x = a_k \\ x & \text{otherwise} . \end{cases}$$

Clearly, $h : X \to X \backslash A$ is one-one and onto.

There are many equivalent forms of AC. We shall use Zorn's lemma—one of its equivalent forms—to prove further results. We explain Zorn's lemma now.

A *partial order* on a set P is a binary relation R such that for $x, y, z \in P$,

i) xRx(reflexive),
ii) xRy and yRz implies xRz (transitive), and
iii) xRy and yRx implies $x = y$ (anti-symmetric).

A set P with a partial order is called *a partially ordered set* or simply a *poset*. A *linear order* on a set X is a partial order R on X such that any two elements of X are comparable, i.e., for every $x, y \in X$, at least one of xRy or yRx holds. If X is a set with more than one element then the inclusion relation $\subset$ on $\mathcal{P}(X)$ is a partial order which is not a linear order. Another example of a partial order which is not a linear order is obtained as follows. Let X and Y be sets with Y having more than one element. Let $\mathcal{P}$ be the set of all one-one partial functions from X to Y (functions with domain a subset of X and range contained in Y). Order $\mathcal{P}$ by 'extension' (if $f, g \in \mathcal{P}$ then g extends f if the domain of f is contained in the domain of g and $f(x) = g(x)$ for every x in the domain of f).

Fix a poset (P, R). A *chain* in P is a subset C of P such that R restricted to C is a linear order, i.e., for any two elements x, y of C, either xRy or yRx must be satisfied. If $A \subset P$, then an *upper bound* for A is an $x \in P$ such that yRx for all $y \in A$. An $x \in P$ is called a *maximal element of* P if for no $y \in P$ different from x, xRy holds. In the second example above, any partial function f with domain X or range Y is a maximal element. So, there may be many maximal elements in a poset which is not linearly ordered.

ZORN'S LEMMA: If P is a non-empty partially ordered set such that every chain in P has an upper bound in P, then P has a maximal element.

Some Consequences of the Axiom of Choice

THEOREM. For any two sets X and Y, at least one of the relations $X \leq_c Y$ or $Y \leq_c X$ holds.

PROOF. Without loss of generality we can assume that both X and Y are non-empty. We need to show that either there exists a one-one map $f : X \to Y$, or there exists a one-one map $g : Y \to X$. Towards this, let P be the set of all one-one partial functions from X to Y. The set P is clearly non-empty. Order it by 'extension'. Let $C = \{f_i : i \in I\}$ be a chain in P. Then C is a consistent family of one-one partial functions. Let $D = \bigcup_{i \in I}$ domain (f_i). Define $f : D \to Y$ by

$$f(x) = f_i(x) \text{ if } x \in \text{ domain } (f_i).$$

Since f_i's are consistent and one-one, f is well-defined and one-one. Thus, f is an upper bound of C. By Zorn's lemma, P has a maximal element, say f_0. We must have domain $(f_0) = X$ or range$(f_0) = Y$. Otherwise, choose $a \in X\setminus$ domain (f_0) and $b \in Y\setminus$ range(f_0). We get a one-one extension of f_0 to domain $(f_0) \bigcup \{a\}$ by assigning b to a. This contradicts the maximality of f_0. If domain $(f_0) = X$, then $X \leq_c Y$ and if range$(f_0) = Y$, then $Y \leq_c X$.

The next result shows that a hotel with infinitely many rooms can, even when full, accommodate as many new guests as are already there!

THEOREM. For every infinite set $X, X \times \{0, 1\} \equiv X$.

PROOF. Let

$$P = \{(A, f) : A \subset X \quad \text{and} \quad f : A \times \{0, 1\} \to A \text{ a bijection}\}.$$

A hotel with infinitely many rooms, even when full, can accommodate as many new guests as already there!

Since X is infinite it contains a countably infinite set, say D. Since $D \times \{0, 1\} \equiv D$, P is non-empty. Partially order P by 'extension'. As before, by Zorn's lemma, get a maximal element (A, f) of P. The result will follow if we show that $A \equiv X$. Since X is infinite, by the theorem proved in the last section, it will be sufficient to show that $X\setminus A$ is finite. Suppose not. Then by AC there is a $B \subset X\setminus A$ such that $B \equiv N$. So, there is a one-one map g from $B \times \{0, 1\}$ onto B. Combining f and g we get a bijection $h : (A \bigcup B) \times \{0, 1\} \to A \bigcup B$ which extends f. This contradicts the maximality of (A, f). Hence, $X\setminus A$ is finite. Therefore, $A \equiv X$. The proof is complete.

COROLLARY. Every infinite set can be written as the union of k-many pairwise disjoint equinumerous sets, where k is any positive integer.

THEOREM. If X is an infinite set then $X \times X \equiv X$.

PROOF. Let

$$P = \{(A, f) : A \subset X \text{ and } f : A \times A \to A \text{ a bijection}\}.$$

Partially order P by 'extension'. By Zorn's lemma take a maximal element (A, f) of P. Note that A must be infinite. To complete the proof, we shall show that $A \equiv X$. Suppose not. Then $A <_c X$.

We first show that $X \backslash A \equiv X$. If possible, suppose $X \backslash A <_c X$. Either $A \leq_c X \backslash A$ or $X \backslash A \leq_c A$. Assume first $X \backslash A \leq_c A$. Take two disjoint sets A_1, A_2 of the same cardinality as A. Now,

$$X = A \bigcup (X \backslash A) \leq_c A_1 \bigcup A_2 \equiv A <_c X.$$

This is a contradiction. Similarly using the other inequality we arrive at the same contradiction. Thus $X \backslash A \equiv X$. Now get $B \subset X \backslash A$ such that $B \equiv A$. Using the last result, write B as the union of three disjoint sets say B_1, B_2 and B_3 each of the same cardinality as A. Since there is a bijection from $A \times A$ to A there exist bijections $f_1 : B \times A \to B_1, f_2 : B \times B \to B_2$ and $f_3 : A \times B \to B_3$. Let $C = A \bigcup B$. Combining these four bijections we get a bijection $g : C \times C \to C$ which extends f. This contradicts the maximality of (A, f).

EXERCISE. Show that an infinite set X is equinumerous to the set of all its finite subsets.

The Continuum Hypothesis (CH)

We now get back to the two famous problems of Cantor mentioned earlier. He conjectured the following. (We are assuming AC.)

THE CONTINUUM HYPOTHESIS (CH). Every uncountable subset of $\mathfrak{R}$ is of the same cardinality as $\mathfrak{R}$.

THE GENERALISED CONTINUUM HYPOTHESIS (GCH). If X is an infinite set then there is no set A such that $X <_c A <_c \mathcal{P}(X)$.

CH says that $\mathfrak{R}$ is the least numerous uncountable set. If GCH were true, then every infinite set is equinumerous to one of the sets obtained by iterating the power set operation on N. Are these statements true? These turned out to be very hard problems and their solutions were startling. In 1938 Kurt Gödel obtained deep results on models of set theory and showed that based on certain axioms for sets (the so called *Zermelo–Fraenkel axioms*), CH and GCH cannot be disproved by producing a 'model' of set theory satisfying the Zermelo–Fraenkel axioms where CH and GCH are satisfied. This was the first time meta-mathematics entered in a non-trivial way to answer a problem in mathematics. Gödel's result does not say that CH and GCH can be proved. In 1963 Paul Cohen developed a very powerful technique, known as *forcing*, to build models of set theory and showed that CH and GCH cannot be proved either. A similar and equally famous result was proved in the nineteenth century. It was shown that the axiom of parallels—through a point outside a given line passes a unique line not intersecting the given line—can neither be proved nor can be disproved. (See K Paranjape's series of articles on Geometry in various issues of *Resonance* in 1996.) Gödel and Cohen's results are generally considered to be among the greatest in twentieth century mathematics. Cohen's method is particularly useful in proving such consistency results. A further discussion on this is beyond the scope of this article.

Arithmetic of Cardinal Numbers

For sets X, Y and Z we can easily check the following.

i) $X \equiv X$,
ii) if $X \equiv Y$ then $Y \equiv X$, and
iii) if $X \equiv Y$ and $Y \equiv Z$ then $X \equiv Z$.

So, to each set X we can assign a symbol, say $|X|$, called its *cardinal number,* such that

$$X \equiv Y \Leftrightarrow |X| \text{ and } |Y| \text{ are the same.}$$

The symbols $|X|$ are called *cardinal numbers* or simply *cardinals*. (Cardinal numbers are denoted by Greek letters κ, λ, μ with or without suffixes.) The cardinal number of a finite set with n elements is denoted by n itself. We put $|N| = \aleph_0$ and $|\mathfrak{R}| = c$.

As in the case of natural numbers, we can add, multiply and compare cardinal numbers. Towards this, let λ and μ be two cardinal numbers and X and Y be two sets such that $|X| = \lambda$ and $|Y| = \mu$.

DEFINITION.

a) $\lambda + \mu = |(X \times \{0\}) \bigcup (Y \times \{1\})|$
b) $\lambda \cdot \mu = |X \times Y|$
c) $\lambda^\mu = |X^Y|$
d) $\lambda \leq \mu$ if $X \leq_c Y$
e) $\lambda < \mu$ if $X <_c Y$.

The above definitions are easily seen to be independent of the choices of X and Y. Further, these extend the corresponding notions for natural numbers. We can extend the notion to that of sum and product for infinite collection of cardinals too. Let $\{\lambda_i\}$ be a set of cardinal numbers. Fix a family $\{X_i\}$ of sets of cardinality λ_i. We define $\prod_i \lambda_i = |\prod_i, X_i|$. To define $\sum_i \lambda_i$ we take X_i's to be pairwise disjoint and put $\sum_i \lambda_i = \left| \bigcup_i X_i \right|$. In terms of these notations note that $\aleph_0 \cdot \aleph_0 = \aleph_0, 2^{\aleph_0} = c, \aleph_0 < c, c^{\aleph_0} = c$ etc.

Cardinal arithmetic is very similar to the arithmetic of natural numbers and follows more or less the same laws. Whatever we have proved about equinumerosity of sets, i.e., the results concerning union, product, $\leq_c$, etc., translate into corresponding results about cardinal numbers. Thus we obtain a very beautiful and useful cardinal arithmetic, remarkably similar to the arithmetic of natural numbers. For instance, the Schröder–Bernstein theorem translates as follows: $\lambda \leq \mu \;\&\; \mu \leq \lambda \Rightarrow \lambda = \mu$; the result on comparability of cardinals becomes: for cardinals λ and μ, at least one of $\lambda \leq \mu$ or $\mu \leq \lambda$ holds. We get some notable differences for infinite cardinals: if λ is infinite, then $\lambda = \lambda + \lambda = \lambda \cdot \lambda$.

Here is how we do cardinal arithmetic.

$$\begin{aligned}
2^c &\leq c^c && (\text{since } 2 \leq c)\\
&= (2^{\aleph_0})^c && (\text{since } c = 2^{\aleph_0})\\
&= 2^{\aleph_0 \cdot c} && (\text{since, for non-empty sets}\\
& && \; X, Y, Z,\ (X^Y)^Z \equiv X^{Y \times Z})\\
&\leq 2^{c \cdot c} && (\text{since } \aleph_0 < c)\\
&= 2^c && (\text{since } c \cdot c = c).
\end{aligned}$$

So, by the Schröder–Bernstein theorem $2^c = c^c$. From this we conclude that there are as many 0–1 valued maps of real numbers as real valued ones!

Suggested Reading

K Kuratowski and A Mostowski. *Set Theory*. North–Holland publishing company. Amsterdam, New York, Oxford, 1976.

Gregory Moore. *Zermelo's Axiom of Choice—its origin, development and influence*. Springer–Verlag. New York, 1982.

Georg Cantor. *Contributions to the Founding of the Theory of Transfinite Numbers*. Dover Publications Inc. New York, 1992.

Nicolas Bourbaki. *Elements of the History of Mathematics*. Springer–Verlag. Berlin, Heidelberg, New York, 1994.

Yiannis N Moschovakis. *Notes on Set Theory*. UTM. Springer–Verlag. New York, Berlin, Heidelberg, 1994.

S M Srivastava
Statistics and Mathematics Unit
Indian Statistical Institute
203 B T Road, Calcutta 700 035
West Bengal
Email: smohan@isical.ac.in

PART II

4

Introductory Remarks

V Pati and A Sitaram

In Part 2 of this volume, we bring together many of the articles from *Resonance* dealing with sequences and series of numbers and functions. Though most students will be familiar with the notion of convergence of a sequence, or series, we record here a few useful facts about convergence.

A sequence $\{a_n\}$ of real or complex numbers is said to *converge* to the finite limit l, if for every $\varepsilon > 0$, there exists a positive integer N (which may depend on ε) such that for all $n \geq N, |a_n - l| < \varepsilon$.

Some situations in which convergence is guaranteed are:

i) If $\{a_n\}$ is an increasing sequence of real numbers which is *bounded above*, then $\{a_n\}$ converges.
ii) If $\{a_n\}$ is a decreasing sequence of real numbers which is *bounded below*, then $\{a_n\}$ converges.
iii) A *Cauchy sequence* is a convergent sequence. (A sequence $\{a_n\}$ of real or complex numbers is said to be a Cauchy sequence if for every $\varepsilon > 0$, there exists a positive integer N, which may depend on ε, such that for all $m, n \geq N, |a_n - a_m| < \varepsilon$.

We finally come to the notion of convergence of an infinite series $\sum_{n=1}^{\infty} a_n$. We consider the sequence of *partial sums* $\{S_n\}$ where $S_n = \sum_{j=1}^{n} a_j$. If this sequence $\{S_n\}$ converges to the finite limit l, then we say that the infinite series $\sum_{n=1}^{\infty} a_n$ is convergent and has the sum l.

5

An Elementary Problem that Interested Ramanujan

Shailesh A Shirali

Ramanujan used to contribute problems/solutions to the Elementary Problems and Solutions section of the *Journal of the Indian Mathematical Society*. The following problem appeared in *JIMS* (Q 289, III 90) and Ramanujan contributed a solution *JIMS* (IV, 226): to determine the value of the "infinite continued square root"

$$\sqrt{1+2\sqrt{1+3\sqrt{1+4\sqrt{1+5\sqrt{1+\cdots}}}}}.$$

We proceed as follows.[1] For $x \geq 0$, define $f(x)$ by

$$f(x) = \sqrt{1+x\sqrt{1+(x+1)\sqrt{1+(x+2)\sqrt{1+\cdots}}}}.$$

To start with, suppose that $x \geq 1$. Observe that

$$\begin{aligned} f(x) &= \sqrt{1+xf(x+1)} \\ &\leq \sqrt{(1+x)f(x+1)} \leq \cdots \leq \prod_{i=1}^{\infty}(x+i)^{1/2^i}. \end{aligned}$$

Using the inequalities $x+i \leq 2xi$ for $x, i \geq 1$ and $\prod i^{1/2^i} < \prod 2^{(i-1)/2^i} = 2$, we obtain

[1] The solution given here is different from that given by Srinivasa Ramanujan and follows that given by D J Newman for the same problem, in his book *A Problem Seminar*.

$$\prod_{i=1}^{\infty}(x+i)^{1/2^i} \le$$

$$\prod_{i=1}^{\infty}(2xi)^{1/2^i} = 2x \cdot \left(\prod_{i=1}^{\infty} i^{1/2^i}\right) < 2x \cdot 2 = 4x.$$

Thus $f(x) < 4x$ for $x \ge 1$. If $0 < x < 1$ then $x+1 > 1$, leading to $f(x+1) < 4(x+1)$, and

$$f(x) < \sqrt{1 + x(4x+4)} = 1 + 2x < 1 + 4x.$$

It follows that $f(x) \le 1 + 4x$ for all $x \ge 0$.

The iterative dance now commences. Suppose that for some $a > 0$ we know that $f(x) \le 1 + ax$ for all $x \ge 0$. Then $f(x+1) \le 1 + a + ax$, and so

$$\begin{aligned} f(x) &\le \sqrt{1 + x(1 + a + ax)} \\ &= \sqrt{1 + (a+1)x + ax^2} \\ &\le 1 + \frac{a+1}{2}x, \end{aligned}$$

since $a \le \left(\frac{a+1}{2}\right)^2$.

Thus $f(x) \le 1 + ax$ implies $f(x) \le 1 + \frac{a+1}{2}x$. Repeating this step, and noting that for any $a > 0$ the sequence $a, \frac{a+1}{2}, \frac{a+3}{4}, \frac{a+7}{8}, \ldots$ converges to 1, we conclude that $f(x) \le 1 + x$ for all $x \ge 0$.

Now for the lower bound. Clearly $f(x+1) \ge f(x)$ for all $x \ge 0$, so it follows that $f(x) \ge \sqrt{1 + xf(x)}$. The inequality is easily solved to give $f(x) \ge (x + \sqrt{x^2+4})/2$. We therefore have,

$$f(x) \ge \frac{x}{2} + \sqrt{1 + \frac{x^2}{4}} \ge 1 + \frac{x}{2} \text{ for all } x \ge 0.$$

The anchor has been secured and the iteration can begin. Assume that for some $a > 0$, we have $f(x) \ge 1 + ax$ for all $x \ge 0$. We now have,

$$\begin{aligned} f(x) &= \sqrt{1 + xf(x+1)} \\ &\ge \sqrt{1 + x(1 + a + ax)} = \sqrt{1 + (1+a)x + ax^2} \\ &\ge 1 + \sqrt{a}x, \text{ since } 1 + a \ge 2\sqrt{a}. \end{aligned}$$

Thus $f(x) \ge 1 + ax$ implies $f(x) \ge 1 + \sqrt{a}x$. Repeating this step, and noting that for any $a > 0$, the sequence $a, a^{1/2}, a^{1/4}, a^{1/8}, \ldots$ converges to 1, we conclude that $f(x) \ge 1 + x$ for all $x \ge 0$.

The sandwich has closed with a pleasing finality about it, and we arrive at the beautiful result: $f(x) = 1 + x$. In particular, the answer to the original problem is $f(2) = 3$.

COROLLARY. There is precisely one function $f : [0, \infty) \to [0, \infty)$ which increases monotonically with x and for which $f(x) = \sqrt{1 + xf(x+1)}$ for all $x \geq 0$; namely, the function $f(x) = 1 + x$.

Notes

1. In the text the function $f(x)$ was defined for $x \geq 0$ by

$$f(x) = \sqrt{1 + x\sqrt{1 + (x+1)\sqrt{1 + (x+2)\sqrt{1 + \cdots}}}}. \qquad (2)$$

However it is not clear that this is a definition at all! It may be better to say that the object on the right side is only a *notation* for a function $f : [0, \infty) \to [0, \infty)$ satisfying the functional equation $f(x) = \sqrt{1 + xf(x+1)}$, assuming that such an object exists. What is shown in this article is that such an f exists, is unique, and is given by $f(x) = 1 + x$.

2. The reference to monotonicity appearing in the statement of the corollary at the end of the paper may be omitted.
3. Readers may wonder about the significance of the curious phrase "the iterative dance". In the text it referred to the process by which we steadily tightened the bounds on the function $f(x)$ by applying some relation over and over again. We offer two more examples of this "dance".

EXAMPLE 1. *Derivation of the power series for the sine and cosine functions*

Let I denote the open interval $(0, \pi/2)$. Starting with the inequality $0 < \cos x < 1$ for $x \in I$, two successive integrations over I yield

$$0 < \sin x < x \text{ and } 1 - \frac{x^2}{2} < \cos x < 1 \quad (x \in I).$$

Further integration yields

$$x - \frac{x^3}{6} < \sin x < x \text{ and } 1 - \frac{x^2}{2} < \cos x < 1 - \frac{x^2}{2} + \frac{x^4}{24} \quad (x \in I);$$

and so we continue, in tandem; it is like a dance between two partners. Ultimately we obtain the familiar series expansions for the sine and cosine functions.

EXAMPLE 2. *Newton–Raphson algorithm for estimating roots of equations.* Here the "dance" occurs as follows. Given the equation $f(x) = 0$ and an initial estimate x_1 of any of its roots, we consider the function g defined by

$$g(x) = x - \frac{f(x)}{f'(x)}.$$

Using g we compute the sequence $\{x_n\}_{n=1}^{\infty}$ via the formula

$$x_{n+1} = g(x_n) \quad (n \in N).$$

It can be shown that under certain fairly reasonable conditions the x_n's converge to some root of the equation $f(x) = 0$.

For another example of this iterative process, please refer to the article *A Route to Square Roots and Cube Roots* in this volume.

SHAILESH A SHIRALI
Rishi Valley School
Rishi Valley 517 352
Andhra Pradesh.

6

The Vacillating Mathematician

The Deterministic Version

K B Athreya

A mathematician starts walking from her home to her office. Halfway through she changes her mind and starts returning home. Again halfway through that she changes her mind and starts walking towards her office. Once again halfway through that she starts returning home and so on. The problem is to determine what happens to this vacillating mathematician.

Identifying the mathematician's home as the point zero and her office as the point one we can formulate a sequence $\{X_n\}_0^\infty$ (X_n denoting the position at the nth change point) of numbers in the interval [0, 1] that satisfies the following simple rule:

$$X_0 = 0, X_1 = \frac{1}{2}, X_2 = \frac{1}{4}, X_3 = \frac{1}{4} + \frac{1}{2} \cdot \frac{3}{4} = \frac{5}{8},$$
$$X_4 = \frac{1}{2} \cdot \frac{5}{8} = \frac{5}{16} \dots$$

$$\left.\begin{aligned} \text{Clearly, } X_{2n+1} &= X_{2n} + \frac{(1 - X_{2n})}{2} \text{ for } n \geq 0 \\ X_{2n} &= \frac{1}{2} X_{2n-1} \text{ for } n \geq 1 \end{aligned}\right] \tag{1}$$

Letting $U_n = X_{2n-1}, V_n = X_{2n}$, we see that

$$\begin{aligned} U_{n+1} &= V_n \left(\frac{1}{2}\right) + \frac{1}{2} \\ &= \left(\frac{1}{4}\right) U_n + \frac{1}{2} \text{ for } n \geq 1 \end{aligned} \tag{2}$$

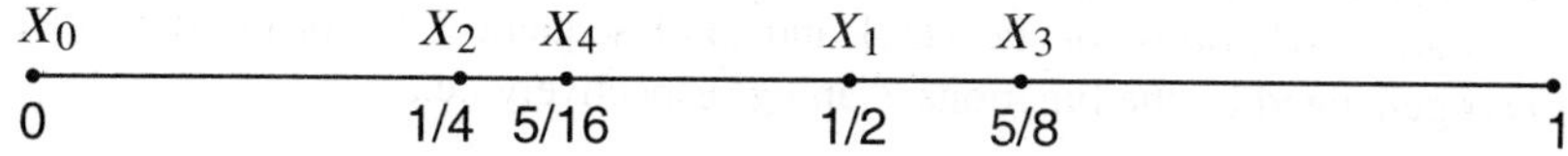

Figure 6.1

This is known as a *first order difference equation*. To solve this, iterate the equation to get

$$\begin{aligned}
U_{n+1} &= \left(\frac{1}{4}\right)\left(\left(\frac{1}{4}\right)U_{n-1}+\frac{1}{2}\right)+\frac{1}{2}\\
&= \left(\frac{1}{4}\right)^2 U_{n-1}+\frac{1}{4}\cdot\frac{1}{2}+\frac{1}{2}\\
&= \left(\frac{1}{4}\right)^2\left(\frac{1}{4}U_{n-2}+\frac{1}{2}\right)+\frac{1}{4}\cdot\frac{1}{2}+\frac{1}{2}\\
&= \left(\frac{1}{4}\right)^3 U_{n-2}+\left(\frac{1}{4}\right)^2\frac{1}{2}+\left(\frac{1}{4}\right)\frac{1}{2}+\frac{1}{2}.
\end{aligned}$$

It is easy to guess from the above and also establish by induction that

$$U_{n+1} = \left(\frac{1}{4}\right)^n U_1 + \sum_{j=0}^{n-1}\left(\frac{1}{2}\right)\left(\frac{1}{4}\right)^j \tag{3}$$

Since $0 \le U_1 \le 1$, $\left(\frac{1}{4}\right)^n \cdot U_1 \to 0$ as $n \to \infty$. Also the geometric series partial sum sequence

$$\sum_0^{n-1}\left(\frac{1}{2}\right)\left(\frac{1}{4}\right)^j \to \frac{\left(\frac{1}{2}\right)}{1-\frac{1}{4}} = \frac{2}{3}.$$

(Recall that $\sum_0^2 ar^j \to \frac{a}{1-r}$ for $|r| < 1$). Thus $U_n \to 2/3$ and since $V_n = X_{2n} = \frac{1}{2}U_n$ it converges to 1/3.

So our mathematician's position at odd numbered change points is nondecreasing (note that $U_2 = 5/8$ is $> U_1 = 1/2$ and $U_{n+1} - U_n = \frac{1}{4}(U_n - U_{n-1}) = (\frac{1}{4})^{n-1}(U_2 - U_1) > 0$) and converges to 2/3. Similarly V_n is also nondecreasing and approaches 1/3.

Eventually, our mathematician will just be hopping in the vicinity of 1/3 and 2/3.

If $0 < X_0 < 1$ then the above arguments are still valid and the limits are the same. However, if $X_0 > 1/3$ then U_n and V_n would both be decreasing to 2/3 and 1/3 respectively. (Prove this).

Finally, note that if $X_0 = 1/3$ then $U_n = 2/3$ and $V_n = 1/3$ for all n. So 2/3 and 1/3 are *fixed points* for the $\{U_n\}$ and $\{V_n\}$ sequences or for the *dynamical systems* generated by the functions f and g respectively i.e.,

$$f(x) = \frac{1}{4}x + \frac{1}{2}, g(x) = \frac{1}{4}x + \frac{1}{4} \tag{4}$$

for x in $[0, 1]$. For a function f from a set S into itself the sequence $\{f_0(x) = x, f_n(x) = f(f_{n-1}(x)), n \geq 1\}$ is called a *dynamical system*. Currently popular topics *chaos* and *fractals* deal with dynamical systems (see Barnsley, and Ramasamy and Iyer in Suggested Reading.)

The problem of the vacillating mathematician was posed by Zeev Barel (Suggested Reading). Krishnapriyan (Suggested Reading) solves this completely and discusses an approach to this problem using difference equations, generating functions and matrix methods.

We conclude this section with the following observation.

Suppose our mathematician when moving towards 1 always goes a fraction α of the remaining distance and when moving towards zero always goes a fraction β of the distance. It can be shown that in this case U_n and V_n satisfy

$$\begin{aligned} U_{n+1} &= (1-\alpha)(1-\beta)U_n + \alpha \\ V_n &= U_n(1-\beta) \end{aligned} \tag{5}$$

and hence

$$U_n \to \frac{\alpha}{1-r} \text{ and } V_n \to \frac{(1-\beta)\alpha}{1-r} \tag{6}$$

where $r = (1-\alpha)(1-\beta)$.

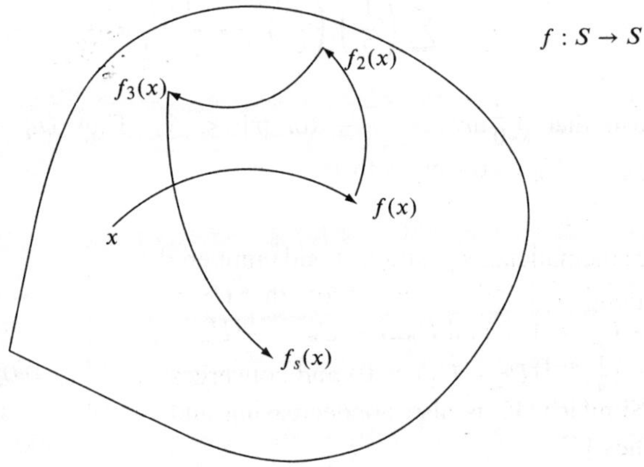

Figure 6.2

Suggested Reading

Zeev Barel. Problem 453. *College Mathematics Journal*. Vol 22. p 255, 1991.
Krishnapriyan H K. *Mathematics Spectrum*. Vol 6. p 9, 1995/6.
Ramaswamy B and TSKV Iyer. *Resonance*. Vol 1. No. 5. May 1996.
Barnsely M. *Fractals Everywhere*, Academic Press, 1988.

K B Athreya
Department of Statistics
Iowa State University
Ames, Iowa, IA 50011, USA

7

A Route to Square Roots and Cube Roots

A Case Study in Serendipity

Shailesh A Shirali

Introduction

Earlier this year, a student of mine reported to me an empirical discovery that he had made: if x is a rational approximation to $\sqrt{2}$, then so is

$$\frac{x+2}{x+1} \quad (= y, \text{say}) \tag{1}$$

and in general it is a better one; for instance, if $x = 4/3$ then $y = 10/7$; and if $x = 3/2$ then $y = 7/5$. Observe that $|y - \sqrt{2}| < |x - \sqrt{2}|$ in each case. The discovery came in response to a question posed earlier to the class, concerning methods for finding rational approximations to square roots. The topic being covered was 'Simplification and Rationalization of Irrational Surds', and I had asked for the best rational approximation to $\sqrt{2}$ with denominator less than 100, with techniques such as the following in mind: raise $(\sqrt{2} - 1)$ to a high positive integral power and express the answer in the form $A\sqrt{2} - B$, where A and B are integers; then $\sqrt{2} \approx B/A$. (For instance, raising it to power 10 yields the approximation

$$\sqrt{2} \approx \frac{3363}{2378} \tag{2}$$

giving 7-place accuracy.) This was duly found (as I had hoped), but observation (1) reported above came as a pleasant surprise. It seems likely that the discovery came about as a result of the following train of ideas. Let $(\sqrt{2} - 1)^k$ be expanded via the binomial theorem and written as $A_k\sqrt{2} - B_k$, where A_k and B_k are integers. Then

$$\begin{aligned}(\sqrt{2}-1)^{k+1} &= (\sqrt{2}-1)(A_k\sqrt{2}-B_k)\\ &= -(A_k+B_k)\sqrt{2}+(2A_k+B_k) \qquad (3)\end{aligned}$$

showing that $A_{k+1} = -(A_k+B_k)$ and $B_{k+1} = -(2A_k+B_k)$. Writing $x = B_k/A_k$ and $y = B_{k+1}/A_{k+1}$, we find that

$$y = \frac{x+2}{x+1} \qquad (4)$$

and since y is clearly better than x as an approximation to $\sqrt{2}$, this suggests observation (1) made above.

The discovery naturally sparked off an investigation into the nature of this and other related approximation schemes, and as the study progressed, an unexpected route to the Newton–Raphson scheme for computing square roots lay revealed. What follows is an account of the investigation.

Brief Resume of Iterative Function Theory

As the proposed scheme is iterative in nature, it is in place to briefly review the basic facts governing such schemes. Let $I \subset \mathbf{R}$ be an open interval (here $\mathbf{R}$ denotes the set of real numbers) and let $f : I \to I$ be a differentiable function. Suppose that f has a fixed point $\alpha \in I$; that is, $f(\alpha) = \alpha$. Suppose moreover that there exists a constant k with $0 < k < 1$ and an open interval $J \subset I$ containing α but no other fixed point of f, such that for all $x \in J, x \neq \alpha$, we have the inequality

$$\left|\frac{f(x)-f(\alpha)}{x-\alpha}\right| < k. \qquad (5)$$

Then f is termed a *contraction mapping* with respect to α (it 'pulls' x towards α), and the fixed point α is termed an *attracting fixed point*. When this happens, the sequence

$$x, f(x), f(f(x)), f(f(f(x))), \ldots \qquad (6)$$

converges to α for all $x \in J$. It can be shown that (5) is equivalent to the requirement that $|f'(\alpha)| < 1$. The value of $|f'(\alpha)|$ in this case gives an approximate indication of the rate of convergence; the smaller the value, the faster the rate of convergence. If the inequality is reversed ($|f'(\alpha)| > 1$), then α is a *repelling fixed point*, and in this case convergence will not take place for any open interval $J \subset I$, however small (unless of course $x = \alpha$). See Smital (Suggested Reading) for further details.

An Iterative Approximation Scheme for Square Roots

Let $a \neq 1$ be a positive rational number, and let a function $f : \mathbf{R}^+ \to \mathbf{R}^+$ (here $\mathbf{R}^+$ denotes the set of positive real numbers) be defined by

$$f(x) = \frac{x+a}{x+1} \tag{7}$$

Our plan is to use the iterates of f to estimate $\sqrt{a}$, which we take to be irrational. We select a starting value $x_0 > 0$ and construct the sequence $\{x_n\}_{n\geq 1}$ via the following rule:

$$x_n = f(x_{n-1}), \quad n > 0 \tag{8}$$

Our hope is that the sequence will converge to $\sqrt{a}$, at least for some restricted set of values for x_0.

Note that in the domain of definition, the only fixed point of f is $x = \sqrt{a}$; for

$$\frac{x+a}{x+1} = x \Rightarrow x^2 = a \Rightarrow x = \sqrt{a} \tag{9}$$

Moreover the fixed point is an attracting one, because

$$f'(x) = -\frac{a-1}{(x+1)^2}, \; \left| f'(\sqrt{a}) \right| = \left| \frac{\sqrt{a}-1}{\sqrt{a}+1} \right| < 1 \tag{10}$$

As per the theory outlined above, this leads us to expect that for any starting number $x_0 > 0$, the sequence $\{x_n\}_{n\geq 1}$ will converge to $\sqrt{a}$.

The conclusion is true and can be proved as follows. We observe firstly that $f(x)$ is monotone throughout its domain: as x goes from 0 to ∞, $f(x)$ goes monotonically from a to 1. Next,

$$f(x) - \sqrt{a} = \frac{(x-\sqrt{a})(1-\sqrt{a})}{x+1}, \quad f(x) - x = \frac{a-x^2}{x+1} \tag{11}$$

So for $a < 1, f(x)$ lies between x and $\sqrt{a}$, whereas for $a > 1, f(x)$ and x lie on opposite sides of $\sqrt{a}$. Now consider the second iterate of f, namely $f_2 = f \circ f$. A quick computation shows that

$$f_2(x) = \frac{x(a+1)+2a}{2x+a+1} \tag{12}$$

It is easy to check the following:

$$f_2(x) - \sqrt{a} = \frac{(\sqrt{a}-1)^2(x-\sqrt{a})}{2x+a+1}$$

$$f_2(x) - x = \frac{2(a-x^2)}{2x+a+1} \tag{13}$$

From this we deduce that $f_2(x)$ always lies between x and $\sqrt{a}$; that is,

$$\begin{aligned} 0 < x < \sqrt{a} &\Rightarrow x < f_2(x) < \sqrt{a} \\ x > \sqrt{a} &\Rightarrow x > f_2(x) > \sqrt{a} \end{aligned} \tag{14}$$

So f_2 is 'better behaved' than f, and it becomes more convenient to work with f_2 than with f. From (13), we obtain, for $x > 0, x \neq \sqrt{a}$:

$$0 < \frac{f_2(x) - \sqrt{a}}{x - \sqrt{a}} = \frac{(\sqrt{a}-1)^2}{2x + (a+1)} < \frac{(\sqrt{a}-1)^2}{a+1} =$$

$$1 - \frac{2\sqrt{a}}{a+1} < 1 \tag{15}$$

So f_2 is a contraction mapping with respect to $\sqrt{a}$: that is, for any $x > 0$, $f_2(x)$ is better than x as an approximation to $\sqrt{a}$. Therefore the sequence

$$\left|x_0 - \sqrt{a}\right|, \left|x_2 - \sqrt{a}\right|, \left|x_4 - \sqrt{a}\right|, \left|x_6 - \sqrt{a}\right|, \ldots \tag{16}$$

decreases monotonically, and as its terms are non-negative it converges to a limit. The limit must be 0, because the only fixed point of f_2 is $\sqrt{a}$. Thus the sequence

$$x_0, x_2, x_4, x_6, \ldots \tag{17}$$

converges to $\sqrt{a}$. We find similarly that the sequence

$$x_1, x_3, x_5, x_7, \ldots \tag{18}$$

converges to $\sqrt{a}$. The assertion made above thus follows: that $x_n \to \sqrt{a}$, as $n \to \infty$, for any value of $x_0 > 0$.

EXAMPLE. The sequence generated with $a = 4$ and $x_0 = 1$ is the following:

$$1, \frac{5}{2}, \frac{13}{7}, \frac{41}{20}, \frac{121}{61}, \frac{365}{182}, \frac{1093}{547}, \ldots \tag{19}$$

and the sequence of absolute errors, $\left|x_n - \sqrt{a}\right|$, is:

$$1, \frac{1}{2}, \frac{1}{7}, \frac{1}{20}, \frac{1}{61}, \frac{1}{182}, \frac{1}{547}, \ldots \tag{20}$$

Note that the sequence is approximately a geometric progression with common ratio 1/3. This is consistent with the fact that

$$\left|f'(\sqrt{a})\right| = \frac{2-1}{2+1} = \frac{1}{3} \tag{21}$$

A function such as $(x+a)/(x+1)$ is termed a *fractional linear function*. Within the class of such functions, can we find one that produces speedier convergence to $\sqrt{a}$ than $(x+a)/(x+1)$? Note that the coefficients we use in such a scheme must be rational, else the procedure lacks meaning. We proceed to take a closer look at this question.

The formula $(x + a)/(x + 1)$ can be stumbled upon in the following heuristic manner:

$$x^2 = a \Rightarrow x^2 + x = x + a \Rightarrow x(x+1) =$$

$$x + a \Rightarrow x = \frac{x+a}{x+1} \tag{22}$$

Following this lead, we try the following idea:

$$x^2 = a \Rightarrow x^2 + kx = kx + a \Rightarrow$$

$$x(x+k) = kx + a \Rightarrow x = \frac{kx+a}{x+k} \tag{23}$$

which holds for any real number k (to avoid complications, we take $k > 0$).

Let $g(x)$ denote the function $(kx + a)/(x + k)$, defined for $x \in \mathbf{R}^+$. The plan is to use g for our iterative scheme, so we first select an appropriate value for k. Now,

$$g'(x) = \frac{k^2 - a}{(x+k)^2} \tag{24}$$

and this yields:

$$\left| g'(\sqrt{a}) \right| = \left| \frac{k^2 - a}{(k + \sqrt{a})^2} \right| = \left| \frac{k - \sqrt{a}}{k + \sqrt{a}} \right| < 1 \tag{25}$$

So convergence to $\sqrt{a}$ works out as earlier, and by choosing k close to $\sqrt{a}$, we can generate fairly fast schemes–the closer the better. (We cannot have $k = \sqrt{a}$ as k must be rational whereas $\sqrt{a}$ is irrational.)

Example. Number crunching helps at this point. Let $a = 10$; possible rational approximations to $\sqrt{a}$ are 3 and 19/6. We obtain the following iterative schemes, corresponding to the choices $k = 1$, $k = 3$ and $k = 19/6$ respectively:

$$x \to \frac{x+10}{x+1}, \quad x \to \frac{3x+10}{x+3}, \quad x \to \frac{19x+60}{6x+19} \tag{26}$$

Of these; the third one provides impressive results. For instance, consider the sequence $\{x_n\}_{n \geq 1}$ generated by this scheme, starting with $x_0 = 3$:

$$x_0 = 3,\ x_1 = \frac{117}{37},\ x_2 = \frac{4443}{1405}$$

$$x_3 = \frac{168717}{53353},\ x_4 = \frac{6406803}{2026009}, \ldots \tag{27}$$

The error term, $e_n = \left| x_n - \sqrt{10} \right|$, takes the following approximate values:

$$e_1 \approx 1.1 \times 10^{-4}, e_2 \approx 8.0 \times 10^{-8},$$

$$e_3 \approx 5.6 \times 10^{-11}, e_4 \approx 3.8 \times 10^{-14}, \ldots \tag{28}$$

Thus we have achieved thirteen-place accuracy within four steps– impressive enough! Observe that the error sequence is approximately a geometric progression with common ratio roughly 1/1400, and that

$$\left| \frac{19 - 6\sqrt{10}}{19 + 6\sqrt{10}} \right| = \left(19 - 6\sqrt{10}\right)^2 \approx \frac{1}{1442} \tag{29}$$

It should be clear now that if $\sqrt{a}$ is irrational, then there is no rational number k that gives a 'best possible' fractional linear approximation scheme $x \to (kx + a)/(x + k)$, because one can select rational numbers k arbitrarily close to $\sqrt{a}$.

Stumbling Upon the Newton–Raphson Scheme

It is at this point that an insight strikes us. We have seen how by choosing k close to $\sqrt{a}$ (the closer the better), we obtain increasingly better schemes of the type $x \to (kx + a)/(x + k)$ for obtaining rational approximations to $\sqrt{a}$, and that by iterating the scheme on some appropriately chosen starting value, we obtain increasingly better rational approximations to $\sqrt{a}$. *Why not then use these approximations themselves as candidates for k?* That is, *for k why not substitute x itself*? With this modification, the scheme changes to the following (it is no longer fractional linear):

$$x \to \frac{x \cdot x + a}{x + x}, \text{ that is, } x \to \frac{x + a/x}{2} \tag{30}$$

and lo and behold we have the Newton–Raphson scheme for computing square roots. On our journey to strange lands we have stumbled upon an old friend....

The Newton–Raphson scheme is known to be tremendously fast in its rate of convergence. Considering the heuristic logic used above, it is not hard to guess that it will beat any fractional linear scheme. Consider the following data, obtained with $a = 10$ and $x_0 = 3$:

$$x_1 = 3, \quad x_2 = \frac{19}{6}, \quad x_3 = \frac{721}{228}, \quad x_4 = \frac{1039681}{328776} \tag{31}$$

$$x_5 = \frac{2161873163521}{683644320912}$$

$$x_6 = \frac{9347391150304592810234881}{2955904621546382351702304}, \ldots \tag{32}$$

The error term $e_n = \left| x_n - \sqrt{10} \right|$ takes the following approximate values:

$$e_1 \approx 1.6 \times 10^{-1}, e_2 \approx 4.4 \times 10^{-3}, e_3 \approx 3.0 \times 10^{-6},$$

$$e_4 \approx 1.5 \times 10^{-12}, e_5 \approx 3.4 \times 10^{-25} \tag{33}$$

Note the rapidity with which the error term shrinks to zero!
(Explanation: With $h(x) = (x + a/x)/2$, we have $h'(\sqrt{a}) = 0$. With fractional linear approximation schemes, using rational coefficients, a derivative of 0 is not possible. A zero value for the derivative implies that the error term in the Newton–Raphson scheme shrinks roughly at a quadratic rate; each term is (roughly) the square of the preceding term. This is best understood in terms of the Taylor series expansion of $h(x)$ about $x = \sqrt{a}$: the linear term is absent, by virtue of the zero derivative, so the rate of convergence is controlled by the second-degree term.)

Cube Roots

Armed with this insight, we now take on the corresponding problem involving cube roots. In this case, no fractional linear scheme will do, for, with rational p, q, r, s, the mapping

$$x \to \frac{px + q}{rx + s} \tag{34}$$

cannot have a fixed point involving an irrational cube root. So we need to examine other types of schemes, and one possibility is to use fractional quadratic schemes.

Let $a > 0$; we need to estimate the value of $a^{1/3}$. Assume that a is rational while $a^{1/3}$ is irrational. Proceeding heuristically as we did earlier, we obtain the following:
For any rational number $k > 0$,

$$x^3 = a \Rightarrow x^3 + kx^2 = kx^2 + a \Rightarrow x = \frac{kx^2 + a}{x^2 + kx} \tag{35}$$

Let f be the function defined thus on $\mathbf{R}^+$:

$$f(x) = \frac{kx^2 + a}{x^2 + kx} \tag{36}$$

The only fixed point of f is $x = a^{1/3}$, as is easily verified. Next,

$$f'(x) = \frac{k^2x^2 - 2ax - ak}{x^2(x + k)^2} \tag{37}$$

We must choose k so that $f'(a^{1/3}) \approx 0$. This is accomplished by ensuring that

$$k^2a^{2/3} - ak - 2a^{4/3} \approx 0 \tag{38}$$

Solving the quadratic equation for k, we find:

$$k \approx 2a^{1/3} \text{ or } -a^{1/3} \tag{39}$$

We discard the negative root. The conclusion then, is that the iterative scheme

$$x \to \frac{kx^2 + a}{x^2 + kx} \tag{40}$$

with $k \approx 2a^{1/3}$ (k rational) should provide us with good rational approximations to $a^{1/3}$.

Numerical experimentation can be attempted at this point. For $a = 2$, we know that 5/4 is close to $a^{1/3}$, so we choose $k = 5/2$. The scheme is therefore:

$$x \to \frac{5x^2 + 4}{x(2x + 5)} \tag{41}$$

With a starting value of $x_0 = 1$, the convergents obtained are the following:

$$x_1 = \frac{9}{7}, \quad x_2 = \frac{601}{477}$$

$$x_3 = \frac{2716121}{2155787}, \quad x_4 = \frac{55476236790681}{44031518284417}, \ldots \tag{42}$$

and $\left|2^{1/3} - x_4\right| \approx 10^{-9}$. This can be considered as fairly satisfactory.

As for the case of square roots, if we limit ourselves to rational values for k, there is no best possible fractional quadratic scheme. Also, just as we did earlier, we can use the updated value of x itself to select k; that is, choose $k = 2x$. The scheme now becomes:

$$x \to \frac{2x \cdot x^2 + a}{x(x + 2x)}, \text{ that is,} x \to \frac{2x^3 + a}{3x^2} \tag{43}$$

and this is exactly the Newton–Raphson scheme for cube roots.

Further Investigations

In terms of pedagogy, one can envisage several investigations in these general directions. Here is one possibility.

Let $x \to f(x), x \to g(x)$ be two iterative schemes that converge (for suitable starting values) to the same fixed point α. Let $f'(\alpha) = \gamma, g'(\alpha) = \delta$. Suppose that γ and δ are non-zero and distinct. Let

$$h(x) = \frac{\gamma g(x) - \delta f(x)}{\gamma - \delta} \tag{44}$$

Then $h(\alpha) = \alpha$ and $h'(\alpha) = 0$. Now let γ', δ' be distinct rational numbers close to γ, δ respectively. It seems reasonable to expect that the scheme

$$x \to \frac{\gamma' g(x) - \delta' f(x)}{\gamma' - \delta'} \tag{45}$$

will converge rapidly. In general, the question of how one can produce a good approximation scheme from two not-so-good approximation schemes, seems a worthwhile one to take up for investigation by student groups.

Conclusion

As every mathematician and every mathematics teacher knows, one of the great pleasures in doing mathematics is the joy of stumbling upon hidden connections serendipitously. The above investigation would seem to bear this out in full measure.

Suggested Reading

Jaroslav Smital. *On Functions and Functional Analysis*. Adam Hilger, 1988.

SHAILESH A SHIRALI
Rishi Valley School
Rishi Valley 517 352
Andhra Pradesh.

8

On the Infinitude of Primes

Euler's Proof

Shailesh A Shirali

Euclid's elegant proof that there are infinitely many prime numbers is well known. Euler proved the same result, in fact a stronger one, by *analytical* methods. This article gives an exposition of Euler's proof introducing the necessary concepts along the way.

Introduction

In this article, we present Euler's very beautiful proof that there are infinitely many prime numbers. In an earlier era, Euclid had proved this result in a simple yet elegant manner. His idea is easy to describe. Denoting the prime numbers by $p_1, p_2, p_3, \ldots$, so that $p_1 = 2, p_2 = 3, p_3 = 5, \ldots$, he supposes that there are n primes in all, the largest being p_n. He then considers the number N where

$$N = p_1 p_2 p_3 \cdots p_n + 1,$$

and asks what the prime factors of N could be. It is clear that N is indivisible by each of the primes $p_1, p_2, p_3, \ldots, p_n$ (indeed, $N \equiv 1 \pmod{p_i}$ for each $i, 1 \le i \le n$). Since every integer greater than 1 has a prime factorization, this forces into existence prime numbers other than the p_i. Thus there can be no largest prime number, and so the number of primes is infinite.

The underlying idea of Euler's proof is very different from that of Euclid's proof. In essence, he proves that the *sum of the reciprocals of the primes is infinite;* that is,

$$\frac{1}{p_1} + \frac{1}{p_2} + \frac{1}{p_3} + \cdots = \infty$$

Euler proved that the sum of the reciprocals of the primes is infinite; the infinitude of the primes thus follows as a corollary.

In technical language, the series $\Sigma_i 1/p_i$ *diverges*. Obviously, this cannot possibly happen if there are only finitely many prime numbers. The infinitude of the primes thus follows as a corollary. Note that Euler's result is stronger than Euclid's.

Convergence and Divergence

A few words are necessary to explain the concepts of convergence and divergence of infinite series. A series $a_1 + a_2 + a_3 + \cdots$ is said to *converge* if the sequence of partial sums,

$$a_1, a_1 + a_2, a_1 + a_2 + a_3, \cdots,$$

approaches some limiting value, say L; we write, in this case, $\sum_1^\infty a_i = L$. If, instead, the sequence of partial sums grows without any bound, we say that the series *diverges*, and we write, in short, $\sum_1^\infty a_i = \infty$.

EXAMPLES.

1) The series $1/1 + 1/2 + 1/4 + \cdots + 1/2^n + \cdots$ converges (the sum is 2, as is easily shown).
2) The series $1/1 + 1/3 + 1/9 + \cdots + 1/3^n + \cdots$ converges (the sum in this case is 3/2).
3) The series $1 + 1 + 1 + \cdots$ diverges (rather trivially).
4) The series $1 - 1 + 1 - 1 + 1 - 1 + 1 - \cdots$ also fails to converge, because the partial sums assume the values $1, 0, 1, 0, 1, 0, \ldots$, and this sequence clearly does not possess a limit.
5) A more interesting example: $1 - 1/2 + 1/3 - 1/4 + \cdots$; a careful analysis shows that it too is convergent, the limiting sum being ln 2 (the natural logarithm of 2).

Divergence of the Harmonic Series $\Sigma 1/i$

In order to prove Euler's result, namely, the divergence of $\Sigma 1/p_i$, we need to establish various subsidiary results. Along the way we shall meet other examples of divergent series. To start with, we present the proof of the statement that

$$\frac{1}{1} + \frac{1}{2} + \frac{1}{3} + \frac{1}{4} \cdots = \infty$$

A statement of the form $\Sigma a_i = \infty$ to be regarded as merely a short form for the statement that the sums $a_1, a_1 + a_2, a_1 + a_2 + a_3, \ldots$, do not possess any limit. It is important to note that ∞ is not to be regarded as a number! We shall however frequently use phrases of the type, $x = \infty$, (for various quantities x) during the course of this article. The meaning should be clear from the context.

This rather non-obvious result is usually referred to as *the divergence of the harmonic series*. The proof given below is due to the Frenchman, Nicolo Oresme, and it dates back to about 1350. We note the following sequence of equalities and inequalities:

$$\frac{1}{1} = \frac{1}{1}$$
$$\frac{1}{2} = \frac{1}{2}$$
$$\frac{1}{3} + \frac{1}{4} > \frac{1}{4} + \frac{1}{4} = \frac{1}{2}$$
$$\frac{1}{5} + \frac{1}{6} + \frac{1}{7} + \frac{1}{8} > \frac{1}{8} + \frac{1}{8} + \frac{1}{8} + \frac{1}{8} = \frac{1}{2}$$
$$\frac{1}{9} + \frac{1}{10} + \cdots + \frac{1}{16} > \frac{1}{16} + \frac{1}{16} + \cdots + \frac{1}{16} = \frac{1}{2}$$

The earliest known proof of the divergence of the harmonic series is due to the Frenchman Nicolo Oresme and it dates to about 1350.

and so on. This shows that it is possible to group consecutive sets of terms of the series $1/1 + 1/2 + 1/3 + \cdots$ in such a manner that each group has a sum exceeding 1/2. Since the number of such groups is infinite, it follows that the sum of the whole series is itself infinite. (Note the crisp and decisive nature of the proof!)

Based on this proof, we make a more precise statement. Let $S(n)$ denote the sum

$$\frac{1}{1} + \frac{1}{2} + \frac{1}{3} + \cdots + \frac{1}{n},$$

e.g., $S(3) = 11/6$. Generalizing from the reasoning just used, we find that

$$S(2^n) > 1 + \frac{n}{2} \tag{1}$$

(Please fill in the details of the proof on your own.) This means that by choosing n to be large enough, the value of $S(2^n)$ can be made to exceed any given bound. For instance, if we wanted the sum to exceed 100, then (1) assures us that a mere 2^{198} terms would suffice! This suggests the extreme slowness of growth of $S(n)$ with n. Nevertheless it does grow without bound; loosely stated, $S(\infty) = \infty$.

The result obtained above, (1) can also be written in the form,

$$S(n) > 1 + \frac{1}{2}\log_2 n.$$

EXERCISE. Write out a proof of the above inequality.

A much more accurate statement can be made, but it involves calculus. We consider the curve Ω whose equation is $y = 1/x, x > 0$. The area of the region enclosed by Ω, the x-axis and the ordinates $x = 1$ and $x = n$ is equal to $\int_1^n \frac{1}{x} dx$, which simplifies to $\ln n$. Now let the region be

The growth of S(n) with n is extremely slow. For the sum to exceed 100, we would require 2^{198} terms!

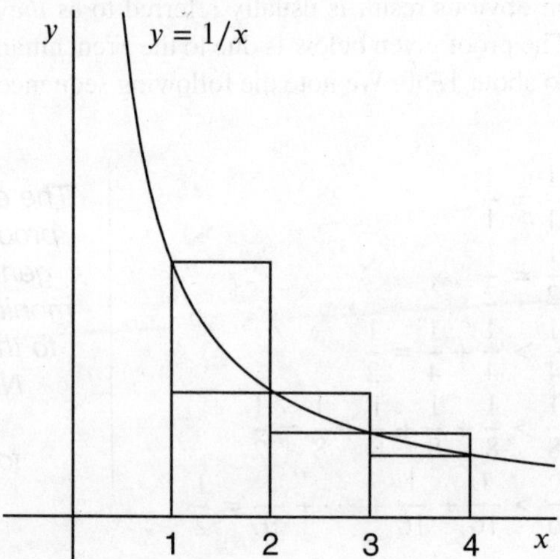

Figure 8.1 *The figure shows how to bound ln n by observing that ln n is the area enclosed by the curve y = 1/x, the x-axis and the ordinates x = 1 and x = n.*

divided into $(n-1)$ strips of unit width by the lines $x = 1, x = 2, x = 3, \ldots, x = n$ (see Figure 8.1).

Consider the region enclosed by Ω, the x-axis, and the lines $x = i-1, x = i$. The area of this region lies between $1/i$ and $1/(i-1)$, because it can be enclosed between two rectangles of dimensions $1 \times 1/i$ and $1 \times 1/(i-1)$, respectively. (A quick examination of the graph will show why this is true.) By letting i take the values $2, 3, 4, \cdots, n$, and adding the inequalities thus obtained, we find that

$$\frac{1}{2} + \frac{1}{3} + \cdots + \frac{1}{n} < \ln n < \frac{1}{1} + \frac{1}{2} + \cdots + \frac{1}{n-1} \tag{2}$$

Relation (3.2) implies that

$$\ln n + \frac{1}{n} < \frac{1}{1} + \frac{1}{2} + \frac{1}{3} + \cdots + \frac{1}{n} < \ln n + 1 \tag{3}$$

In general, when Mathematicians find that a series Σa_i diverges, they are also curious to know how fast it diverges.

and this means that we have an estimate for $S(n)$ (namely, $\ln n + 0.5$) that differs from the actual value by no more than 0.5. A still deeper analysis shows that for large values of n, an excellent approximation for $S(n)$ is $\ln n + 0.577$, but we shall not prove this result here. It is instructive, however, to check the accuracy of this estimate. Write $f(n)$ for $\ln n + 0.577$. We now find the following

n	=	10	100	1000	10000	100000
$S(n)$	=	2.92897	5.18738	7.48547	9.78761	12.0902
$f(n)$	=	2.87959	5.18217	7.48476	9.78734	12.0899

The closeness of the values of $f(n)$ and $S(n)$ for large values of n is striking. (The constant 0.577 is related to what is known as the Euler–Mascheroni constant.)

In general, when mathematicians find that a series Σa_i, diverges, they are also curious to know how *fast* it diverges. That is, they wish to find a function, say $f(n)$, such that the ratio $(\Sigma_1^n a_i)/f(n)$ tends to 1 as $n \to \infty$. For the harmonic series $\Sigma 1/i$, we see that one such function is given by $f(n) = \ln n$. This is usually expressed by saying that the harmonic series diverges like the logarithmic function. We note in passing that this is a very slow rate of divergence, because $\ln n$ diverges more slowly than n^ε for any $\varepsilon > 0$, *no matter how small ε is*, in the sense that $\ln n/n^\varepsilon \to 0$ as $n \to \infty$ for every $\varepsilon > 0$. Obviously the function $\ln \ln n$ diverges still more slowly.

EXERCISE. Prove that if $a > 1$ then the series

$$\frac{1}{1^a} + \frac{1}{2^a} + \frac{1}{3^a} + \cdots$$

converges. (The conclusion holds no matter how close a is to 1, but it does not hold for $a = 1$ or $a < 1$, a curious state of affairs!) Further, use the methods of integral calculus (and the fact that for $a \neq 1$, the integral of $1/x^a$ is $x^{(1-a)}/(1-a)$, to show that the sum of the series lies between $1/(a-1)$ and $a/(a-1)$.

The fact that the sum $1/1 + 1/2^2 + 1/3^2 + \cdots$ is finite can be shown in another manner that is both elegant and elementary. We start with the inequalities, $2^2 > 1 \times 2, 3^2 > 2 \times 3, 4^2 > 3 \times 4, \cdots$, and deduce from these that

The fact that the sum $1/1^2 + 1/2^2 + 1/3^2 + \ldots$ is finite can be shown in a manner that is both elegant and elementary.

$$1+\frac{1}{2^2}+\frac{1}{3^2}+\frac{1}{4^2}+\cdots < 1+\frac{1}{1\times 2}+\frac{1}{2\times 3}+\frac{1}{3\times 4}+\cdots$$

The sum on the right side can be written in the form

$$1+\left(\frac{1}{1}-\frac{1}{2}\right)+\left(\frac{1}{2}-\frac{1}{3}\right)+\left(\frac{1}{3}-\frac{1}{4}\right)+\cdots \tag{4}$$

which (after a whole feast of cancellations) simplifies to $1+1/1$, that is, to 2. (This is sometimes described by stating that the series 'telescopes' to 2.) Therefore the sum $1 + 1/2^2 + 1/3^2 + 1/4^2 + \cdots$ is less than 2. We now call upon a theorem of analysis which states that if the partial sums of any series form an increasing sequence and are at the same time bounded, that is, they do not exceed some fixed number, then they possess a limit. We conclude, therefore, that the series $\Sigma 1/i^2$ does possess a finite sum which lies between 1 and 2.

The divergence of the harmonic series was independently proved by Johann Bernoulli in 1689 in a completely different manner. His proof is worthy of deep study, as it shows the counter-intuitive nature of infinity.

Bernoulli starts by assuming that the series $1/2 + 1/3 + 1/4 + \cdots$ (note that he starts with 1/2 rather 1/1) does have a finite sum, which he calls S. He now proceeds to derive a contradiction in the following manner. He rewrites each term occurring in S thus:

$$\frac{1}{3} = \frac{2}{6} = \frac{1}{6} + \frac{1}{6}, \quad \frac{1}{4} = \frac{3}{12} = \frac{1}{12} + \frac{1}{12} + \frac{1}{12}, \ldots,$$

and more generally,

$$\frac{1}{n} = \frac{n-1}{n(n-1)} = \frac{1}{n(n-1)} + \frac{1}{n(n-1)} + \cdots + \frac{1}{n(n-1)},$$

with $(n-1)$ fractions on the right side. Next he writes the resulting fractions in an array as shown below:

The divergence of the harmonic series was independently proved by Johann Bernoulli in 1689 in a completely different manner.

$$\begin{array}{cccccccc}
1/2 & 1/6 & 1/12 & 1/20 & 1/30 & 1/42 & 1/56 & \cdots \\
 & 1/6 & 1/12 & 1/20 & 1/30 & 1/42 & 1/56 & \cdots \\
 & & 1/12 & 1/20 & 1/30 & 1/42 & 1/56 & \cdots \\
 & & & 1/20 & 1/30 & 1/42 & 1/56 & \cdots \\
 & & & & 1/30 & 1/42 & 1/56 & \cdots \\
 & & & & & 1/42 & 1/56 & \cdots \\
 & & & & & & 1/56 & \cdots
\end{array}$$

Note that the column sums are just the fractions $1/2, 1/3, 1/4, 1/5, \ldots$; thus S is the sum of all the fractions occurring in the array. Bernoulli now sums the rows using the telescoping technique used above (see equation (4)). Assigning symbols to the row sums as shown below,

$$A = \frac{1}{2} + \frac{1}{6} + \frac{1}{12} + \frac{1}{20} + \frac{1}{30} + \frac{1}{42} + \frac{1}{56} + \cdots$$

$$B = \frac{1}{6} + \frac{1}{12} + \frac{1}{20} + \frac{1}{30} + \frac{1}{42} + \frac{1}{56} + \cdots$$

$$C = \frac{1}{12} + \frac{1}{20} + \frac{1}{30} + \frac{1}{42} + \frac{1}{56} + \cdots$$

$$D = \frac{1}{20} + \frac{1}{30} + \frac{1}{42} + \frac{1}{56} + \cdots$$

he finds that:

Bernoulli's proof is worthy of deep study, as it shows the counter-intuitive nature of infinity.

$$\begin{aligned} A &= \left(1 - \frac{1}{2}\right) + \left(\frac{1}{2} - \frac{1}{3}\right) + \left(\frac{1}{3} - \frac{1}{4}\right) \\ &\quad + \left(\frac{1}{4} - \frac{1}{5}\right) + \cdots \\ &= 1 \end{aligned}$$

$$
\begin{aligned}
B &= \left(\frac{1}{2}-\frac{1}{3}\right)+\left(\frac{1}{3}-\frac{1}{4}\right)+\left(\frac{1}{4}-\frac{1}{5}\right) \\
&\quad+\left(\frac{1}{5}-\frac{1}{6}\right)+\cdots \\
&= \frac{1}{2} \\
C &= \frac{1}{3} \qquad \text{(arguing likewise)} \\
D &= \frac{1}{4}
\end{aligned}
$$

and so on. Thus the sum S, which we had written in the form $A+B+C+D+\cdots$, turns out to be equal to

$$1+\frac{1}{2}+\frac{1}{3}+\frac{1}{4}+\cdots$$

Now this looks disappointing—just as things were beginning to look promising! We seem to have just recovered the original series after a series of very complicated steps. But in fact something significant has happened: *an extra '1' has entered the series*. At the start we had defined S to be $1/2+1/3+1/4+\cdots$; now we find that S equals $1 + 1/2 + 1/3 + 1/4 + \cdots$. This means that $S = S + 1$. However, no finite number can satisfy such an equation. Conclusion: $S = \infty$!

There are many other proofs of this beautiful result, but I shall leave you with the pleasant task of coming up with them on your own. Along the way you could set yourself the task of proving that each of the following sums diverge:

- $1/1 + 1/3 + 1/5 + 1/7 + 1/9 + \cdots$;
- $1/1 + 1/11 + 1/21 + 1/31 + 1/41 + \cdots$;
- $1/a + 1/b + 1/c + 1/d + \cdots$, where $a, b, c, d, \cdots$, are the successive terms of any increasing arithmetic progression of positive real numbers.

Elementary Results

The next result that we shall need is the so-called fundamental theorem of arithmetic: *every positive integer greater than* 1 *can be expressed in precisely one way as a product of prime numbers*. We shall not prove this very basic theorem of number theory. For a proof, please refer to any of the well-known texts on number theory, e.g., the text by Hardy and Wright, or the one by Niven and Zuckermann.

The fundamental theorem of arithmetic states that every positive integer greater than one can be expressed in precisely one way as a product of prime numbers.

We shall also need the following rather elementary results: (i) if k is any integer greater than 1, then

$$\frac{1}{1-1/k} = 1 + \frac{1}{k} + \frac{1}{k^2} + \frac{1}{k^3} + \frac{1}{k^4} + \cdots \tag{5}$$

which follows by summing the geometric series on the right side, and (ii) if a_i, b_j are any quantities, then

$$\left(\sum_i a_i\right)\left(\sum_j b_j\right) = \sum_{i,j} a_i b_j$$

where, in the sum on the right, each pair of indices (i, j) occurs *precisely once*.

Now consider the following two equalities, which are obtained from (5) using the values $k = 2$, $k = 3$:

$$\frac{1}{1-1/2} = 1 + \frac{1}{2} + \frac{1}{2^2} + \frac{1}{2^3} + \frac{1}{2^4} + \cdots$$

$$\frac{1}{1-1/3} = 1 + \frac{1}{3} + \frac{1}{3^2} + \frac{1}{3^3} + \frac{1}{3^4} + \cdots$$

We multiply together the corresponding sides of these two equations. On the left side we obtain $2 \times 3/2 = 3$. On the right side we obtain the product

$$(1 + 1/2 + 1/2^2 + 1/2^3 + \cdots) \times (1 + 1/3 + 1/3^2 + 1/3^3 + \cdots)$$

Expanding the product, we obtain:

$$1 + \frac{1}{2} + \frac{1}{4} + \frac{1}{8} + \cdots + \frac{1}{3} + \frac{1}{9} + \frac{1}{27} + \cdots$$
$$+ \frac{1}{6} + \frac{1}{12} + \frac{1}{24} + \cdots + \frac{1}{18} + \frac{1}{36} + \frac{1}{72} + \cdots,$$

We obtain a nice corollary: if A denotes the set of integers of the form 2^a3^b, where a and b are non-negative integers, then $\sum_{z \varepsilon A} 1/Z = 3$.

that is, we obtain the sum of the reciprocals of all the positive integers that have only 2 and 3 among their prime factors. The fundamental theorem of arithmetic assures us that each such integer occurs *precisely once* in the sum on the right side. Thus we obtain a nice corollary: if A denotes the set of integers of the form $2^a\,3^b$, where a and b are non-negative integers, then

$$\sum_{z \in A} \frac{1}{z} = 3$$

If we multiply the left side of this relation by $(1 + 1/5 + 1/5^2 + 1/5^3 + \cdots)$ and the right side by $\frac{1}{1-1/5}$, we obtain the following result:

$$\sum_{z\in B} \frac{1}{z} = \frac{3}{1-1/5} = \frac{15}{4}$$

where B denotes the set of integers of the form $2^a\,3^b\,5^c$, where a, b, and c denote non-negative integers.

Continuing this line of argument, we see that infinitely many such statements can be made, for example:

1) If C denotes the set of positive integers of the form $2^a\,3^b\,5^c\,7^d$, where a, b, c and d are non-negative integers, we then have $\Sigma_{z\in C} 1/z = (15/4)(7/6) = 35/8$.
2) If D denotes the set of positive integers of the form $2^a\,3^b\,5^c\,7^d\,11^e$, then $\Sigma_{z\in D} 1/z = (35/8)(11/10) = 77/16$.

Infinitude of the Primes

Suppose now that there are only finitely many primes, say $p_1, p_2, p_3, \cdots, p_n$, where $p_1 = 2, p_2 = 3, p_3 = 5, \cdots$. We consider the product

$$\frac{1}{1-1/2}\,\frac{1}{1-1/3}\,\frac{1}{1-1/5}\cdots\frac{1}{1-1/p_n}$$

This is obviously a finite number, being the product of finitely many non-zero fractions. Now this product also equals

$$\left(1+\frac{1}{2}+\frac{1}{2^2}+\cdots\right)\times\left(1+\frac{1}{3}+\frac{1}{3^2}+\cdots\right)\times$$

$$\left(1+\frac{1}{5}+\frac{1}{5^2}+\cdots\right)\times\cdots\times\left(1+\frac{1}{p_n}+\frac{1}{p_n^2}+\cdots\right)$$

When we expand out this product, we find, by continuing the line of argument developed above, that we obtain *the sum of the reciprocals of all the positive integers*. To see why, we need to use the fundamental theorem of arithmetic and the assumption that $2, 3, 5, \cdots, p_n$ are *all* the primes that exist; these two statements together imply that every positive integer can be expressed *uniquely* as a product of non-negative powers of the n primes $2, 3, 5, \cdots, p_n$. From this it follows that the expression on the right side is precisely the sum

$$\frac{1}{1}+\frac{1}{2}+\frac{1}{3}+\frac{1}{4}+\cdots$$

written in some permuted order. But by the Oresme–Bernoulli theorem, the latter sum is infinite! So we have a contradiction: the finite number

$$\frac{1}{1-1/2}\,\frac{1}{1-1/3}\,\frac{1}{1-1/5}\cdots\frac{1}{1-1/p_n}$$

has been shown to be infinite—an absurdity! The only way out of this contradiction is to drop the assumption that there are only finitely many prime numbers. Thus we reach the desired objective, namely, that of proving that there are infinitely many prime numbers.

Note that, as a bonus, there are several formulas that drop out of this analysis, more or less as corollaries. For instance, we find that

$$\frac{1}{1-1/2^2}\frac{1}{1-1/3^2}\frac{1}{1-1/5^2}\cdots = 1+\frac{1}{2^2}+\frac{1}{3^2}+\frac{1}{4^2}+\cdots$$

Euler was capable of stunning reasoning; some of the steps in his proofs are so daring that they would leave today's mathematicians gasping for breath.

that is, the infinite product and the infinite sum both converge to the same (finite) value. By a stunning piece of reasoning, including a few daring leaps that would leave today's mathematicians gasping for breath, Euler showed that both sides of the above equation are equal to $\pi^2/6$. Likewise, we find that

$$\frac{1}{1-1/2^4}\frac{1}{1-1/3^4}\frac{1}{1-1/5^4}\cdots = 1+\frac{1}{2^4}+\frac{1}{3^4}+\frac{1}{4^4}+\cdots,$$

It is not for nothing that Euler is at times referred to as analysis incarnate!

and this time both sides converge to $\pi^4/90$. Euler proved all this and much much more; it is not for nothing that he is at times referred to as *analysis incarnate!*

The Divergence of $\sum 1/p$

As mentioned earlier, Euler showed in addition that the sum

$$\sum_{i\geq 1}\frac{1}{p_i} = \frac{1}{2}+\frac{1}{3}+\frac{1}{5}+\frac{1}{7}+\cdots$$

is itself infinite. We are now in a position to obtain this beautiful result. For any positive integer $n \geq 2$, let P_n denote the set of prime numbers less than or equal to n. We start by showing that

$$\prod_{p\in P_n}\frac{1}{1-1/p} > \sum_{j=1}^{n}\frac{1}{j}. \tag{6}$$

Our strategy will be a familiar one. We write down the following inequality for each $p \in P_n$, which follows from equation (5)

$$\frac{1}{1-1/p} > 1+\frac{1}{p}+\frac{1}{p^2}+\frac{1}{p^3}+\cdots+\frac{1}{p^n}$$

Leonhard Euler

Since universities were not the major research centres in his days, Leonhard Euler (1707–1783) spent most of his life with the Berlin and Petersburg Academies. Pious, but not dogmatic, Euler conducted prayers for his large household, and created mathematics with a baby on his lap and children playing all around. Euler withheld his own work on calculus of variations so that young Lagrange (1736–1813) could publish it first, and showed similar generosity on many other occasions. Utterly free of false pride, Euler always explained how he was led to his results saying that "the path I followed will perhaps be of some help". And, indeed, generations of mathematicians followed Laplace's advice: "Read Euler, he is our master in all!".

The '>' sign holds because we have left out all the positive terms that follow the term $1/p^n$. Multiplying together the corresponding sides of all these inequalities ($p \in P_n$), we obtain:

$$\prod_{p \in P_n} \frac{1}{1 - 1/p} > \prod_{p \in P_n} \left(1 + \frac{1}{p} + \frac{1}{p^2} + \frac{1}{p^3} + \cdots + \frac{1}{p^n}\right)$$

When we expand out the product on the right side, we obtain a sum of the form $\Sigma_{j \in A} 1/j$ for some set of positive integers A. This set certainly includes all the integers from 1 to n because the set P_n contains all the prime numbers between 1 and n. Inequality (6) thus follows immediately.

Next, we already know (see equation (3)) that

$$\sum_{j=1}^{n} \frac{1}{j} > \ln n + \frac{1}{n} > \ln n \tag{7}$$

Combining (6) and (7), we obtain the following inequality:

$$\prod_{p \in P_n} \frac{1}{1 - 1/p} > \ln n$$

Taking logarithms on both sides, this translates into the statement,

$$\sum_{p \in P_n} \ln \left(\frac{1}{1 - 1/p}\right) > \ln \ln n \tag{8}$$

Our task is nearly over. It only remains to relate the sum $\Sigma_{p \in P_n} 1/p$ with the sum on the left side of (8). We accomplish this by showing that the inequality

$$\frac{7x}{5} > \ln \left(\frac{1}{1 - x}\right) \tag{9}$$

holds for $0 < x \leq 1/2$.

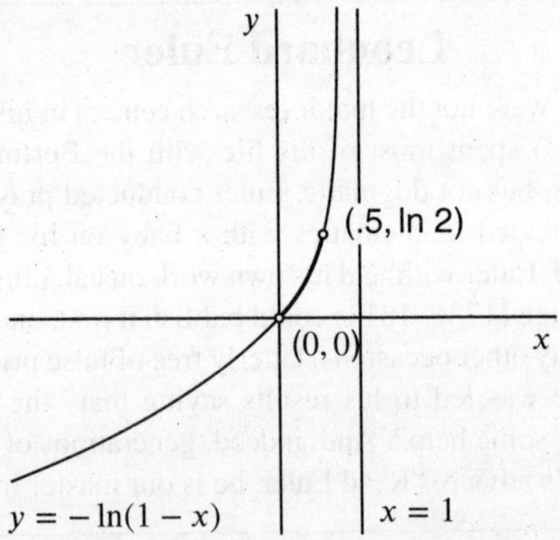

Figure 8.2 *The graph shows that for* $0 \le x \le 1/2$, *we have* $(2\,ln\,2)\,x \ge ln\,(1/(1-x))$.

To see why (9) is true, draw the graph of the curve Γ whose equation is $y = \ln(1/(1-x))$, over the domain $-\infty < x < 1$, (see Figure 8.2). Note that Γ passes through the origin and is convex over its entire extent. (Proof: Write $f(x) = -\ln(1-x)$; then $f'(x) = 1/(1-x)$ and $f''(x) = 1/(1-x)^2 > 0$ for all $x < 1$.)

Euler's Prodigious Output

It has been estimated that Euler's 886 works would fill 80 large books. Dictating or writing on his slate, Euler kept up his unparalleled output all through his life. Though totally blind for the last 17 years of his life, he promised to supply the Petersburg Academy with papers until 20 years after his death; one came out 79 years after he died! The most prolific mathematician in history died while playing with his grandchildren and drinking tea. (*All boxed notes on Euler taken from the IBM poster* Men of Modern Mathematics 1966.)

The convexity of Γ implies that the chord joining the points $A(0,0)$ and $B(1/2, \ln 2)$ lies completely *above* the curve. The equation of AB is $y = (2\ln 2)\,x$, so over the range $0 \le x \le 1/2$ we have the inequality:

$$(2\ln 2)\,x \ge \ln\left(\frac{1}{1-x}\right)$$

Since $\ln 2 \approx 0.69315 < 0.7 = 7/10$, (9) follows. Inequality (9) implies that

$$x > \frac{5}{7}\ln\left(\frac{1}{1-x}\right)$$

for $x = 1/2, x = 1/3, x = 1/5, \cdots$. Therefore, by addition,

$$\sum_{p \in P_n} \frac{1}{p} > \frac{5}{7} \left(\sum_{p \in P_n} \ln \frac{1}{1 - 1/p} \right) \tag{10}$$

Combining (8) and (10), we deduce that

$$\sum_{p \in P_n} \frac{1}{p} > \frac{5}{7} \ln \ln n$$

As $n \to \infty$, the right side diverges to infinity, therefore so does the left side; so we reach our desired objective, that of showing the divergence of $\Sigma_i 1/p_i$.

The Genius of Euler

Euler's work on the zeta function, partitions and divisor sums remind us that he founded analytic number theory, while his creation of the theory of residues of powers and his proof of Fermat theorems are permanent contributions to elementary number theory. Various Euler equations establish his claims to mechanics, calculus of variations and hydrodynamics (where he gave the Lagrangian form as well as the Eulerian). The systematic theory of continued fractions is his, as is a major method in divergent series—justified, a century later, by analytic continuation. Analytic trigonometry, quadric surfaces, theory of investment and annuities, and linear differential equations with constant coefficients are among the many elementary subjects whose present form is chiefly due to Euler.

An Alternative Proof

Here is an alternative proof of the claim that $\Sigma_i 1/p_i$ diverges. The proof has been written in an 'old-fashioned' style and purists will protest. Nevertheless, we shall present the proof and let readers judge for themselves. Let S denote the sum $\Sigma_i 1/p_i$. We shall make use of the following result:

$$e^x \geq 1 + x \text{ for all real values of } x,$$

with equality holding precisely when $x = 0$. The graphs of e^x and $1+x$ show why this is true; the former graph is convex over its entire extent (examine the second derivative of e^x to see why), while the latter, a line, is tangent to the former at the point $(0, 1)$, and lies entirely below it everywhere else. Substituting the values $x = 1/2, x = 1/3, x = 1/5, \cdots$, successively into this inequality, we find that

$$e^{1/2} > 1 + \frac{1}{2}, \quad e^{1/3} > 1 + \frac{1}{3}, \quad e^{1/5} > 1 + \frac{1}{5}, \cdots$$

Multiplying together the corresponding sides of these inequalities, we obtain:

$$e^S > \left(1+\frac{1}{2}\right)\left(1+\frac{1}{3}\right)\left(1+\frac{1}{5}\right)\cdots$$

The infinite product on the right side yields the following series:

$$1+\frac{1}{2}+\frac{1}{3}+\frac{1}{5}+\frac{1}{6}+\frac{1}{7}+\frac{1}{10}+\frac{1}{11}+\frac{1}{13}+\frac{1}{14}+\frac{1}{15}+\cdots$$

Johann Bernoulli

Johann Bernoulli (1667–1748), the younger brother of Jakob Bernoulli (1654–1705) took it upon himself to spread Leibniz's calculus across the European continent. Johann's proof of the divergence of the harmonic series first appeared (1689) in Jakob's treatise and with uncharacteristic fraternal affection, Jakob even prefaced the argument with an acknowledgement of his brother's priority.

This series is the sum of the reciprocals of all the positive integers whose prime factors are all distinct; equivalently, the positive integers that have no squared factors. These numbers are sometimes referred to as the *quadratfrei* or *square-free* numbers. Let Q denote this sum. We shall show that this series itself diverges, in other words, that $Q = \infty$. This will immediately imply that $S = \infty$ (for $e^S > Q$), and Euler's result will then follow.

We consider the product

$$Q \times \left(1+\frac{1}{2^2}+\frac{1}{3^2}+\frac{1}{4^2}+\cdots\right)$$

This product, when expanded out, gives the following series:

$$\frac{1}{1}+\frac{1}{2}+\frac{1}{3}+\frac{1}{4}+\cdots$$

that is, we obtain the harmonic series. To see why, note that every positive integer n can be *uniquely* written as a product of a square-free number and a square; for example, $1000 = 10 \times 10^2, 2000 = 5 \times 20^2, 1728 = 3 \times 24^2$, and so on. Now when we multiply

$$\left(1+\frac{1}{2}+\frac{1}{3}+\frac{1}{5}+\frac{1}{6}+\frac{1}{7}+\frac{1}{10}+\frac{1}{11}+\frac{1}{13}+\frac{1}{14}+\frac{1}{15}+\cdots\right)$$

with

$$\left(1+\frac{1}{2^2}+\frac{1}{3^2}+\frac{1}{4^2}+\cdots\right)$$

we find, by virtue of the remark just made, that the reciprocal of each positive integer n occurs *precisely once* in the expanded product. This explains why the product is just the harmonic series. Now recall that the sum

$$1 + \frac{1}{2^2} + \frac{1}{3^2} + \frac{1}{4^2} + \cdots$$

is finite (indeed, we have shown that it is less than 2). It follows that

$$Q \times \text{(some finite number)} = \infty.$$

Therefore $Q = \infty$, and Euler's result $(\Sigma_i 1/p_i = \infty)$ follows. QED!

XVI. *Summa seriei infinita harmonicè progressionalium,* $\frac{1}{1}+\frac{1}{2}+\frac{1}{3}+\frac{1}{4}+\frac{1}{5}$*& c, est infinita.*

Id primus deprehendit Frater: inventa namque per præced. Summa seriei $\frac{1}{2} + \frac{1}{6} + \frac{1}{12} + \frac{1}{20} + \frac{1}{30}$, & c. visurus porrò, quid emergeret ex ista serie, $\frac{1}{2} + \frac{2}{6} + \frac{3}{12} + \frac{4}{20} + \frac{5}{30}$& c. Si resolveretur methodo Prop. XIV. collegit p opositionis veritatem ex absurditate manifesta, quæ sequeretur, si summa Seriei harmonicæ finita statueretur. Animadvertit enim, Seriem A, $\frac{1}{2} + \frac{1}{3} + \frac{1}{4} + \frac{1}{5} + \frac{1}{6} + \frac{1}{7}$, & c.∞. (fractionibus singulis in alias, quarum numeratores sunt 1, 2, 3, 4, & c. transmutatis)

$$\text{Seriei B, } \underline{\tfrac{1}{2} + \tfrac{2}{6} + \tfrac{3}{12} + \tfrac{4}{20} + \tfrac{5}{30} + \tfrac{6}{42}, \text{\& c.}\infty \text{C} + \text{D} + \text{E} + \text{F}, \text{\& c.}}$$

$$\left.\begin{array}{lllll}
\text{C}\cdot & \frac{1}{2}+\frac{1}{6}+\frac{1}{12}+\frac{1}{20}+\frac{1}{30}+\frac{1}{42}, & \text{\& c.}\infty & \text{per præc.}\,\frac{1}{1} \\
\text{D}\cdot\cdot & +\frac{1}{6}+\frac{1}{12}+\frac{1}{20}+\frac{1}{30}+\frac{1}{42}, & \text{\& c.}\infty & \text{C}-\frac{1}{2}\infty\frac{1}{2} \\
\text{E}\cdots & +\frac{1}{12}+\frac{1}{20}+\frac{1}{30}+\frac{1}{42}, & \text{\& c.}\infty & \text{D}-\frac{1}{6}\infty\frac{1}{3} \\
\text{F}\cdot\cdot\cdot\cdot & +\frac{1}{20}+\frac{1}{30}+\frac{1}{42}, & \text{\& c.}\infty & \text{E}-\frac{1}{1}\infty\frac{1}{4} \\
 & & \text{\& c.}\infty & \text{\&c.}
\end{array}\right\} \infty\text{G; unde}$$

sequitur, seriem G ∞ A, totum parti, si summa finita effet. Ego

Johann's divergence proof, from Jakob's *Tractatus de Seriebus Infinitis*, republished in 1713. (From page 197 of *Journey through Genius* by William Dunham).

Readers who are unhappy with this style of presentation, in which ∞ is treated as an ordinary real number, will find it an interesting (but routine) exercise to rewrite the proof to accord with more exacting standards of rigour and precision.

Conclusion

A much deeper, but also more difficult analysis shows that the sum $1/p_1 + 1/p_2 + 1/p_3 + \cdots + 1/p_n$ is approximately equal to $\ln \ln n$. This is usually stated in the following form: as n tends to ∞, the fraction

$$\frac{1/p_1 + 1/p_2 + 1/p_3 + \cdots + 1/p_n}{\ln \ln n}$$

tends to 1. This is indeed a striking result, reminiscent of the earlier result that $1/1+1/2+1/3+\cdots+1/n$ is approximately equal to $\ln n$. It shows the staggeringly slow rate of divergence of the sum of the reciprocals of the primes. The harmonic series $\Sigma_i 1/i$, diverges slowly enough—to achieve a sum of over 100, for instance, we would need to add more than 10^{43} terms, so it is certainly not a job that one can leave to finish off over a weekend. (Do you see where the number 10^{43} comes from?) On the other hand, to achieve a sum of over 100 with the series $\Sigma_i 1/p_i$, we need to add something like $10^{10^{43}}$ terms!! This number is so stupendously large that it is a hopeless task to make any visual image of it. Certainly there is no magnitude even remotely comparable to it in the whole of the known universe.

Suggested Reading

G H Hardy, E M Wright. *An Introduction to the Theory of Numbers.* 4th ed., Oxford, Clarendon Press, 1960.

Ivan Niven, Herbert S Zuckermann. *An Introduction to the Theory of Numbers.* Wiley Eastern Ltd., 1989.

Tom Apostol. *An Introduction to Analytic Number Theory.* Narosa Publishing House, 1979.

SHAILESH A SHIRALI
Rishi Valley School
Rishi Valley 517 352
Andhra Pradesh

9

Nīlakaṇṭha, Euler and π

Shailesh A Shirali

Introduction

In this article, we present brief expository accounts of two well-known series for π: the *Gregory–Nīlakaṇṭha series,*

$$\frac{\pi}{4} = 1 - \frac{1}{3} + \frac{1}{5} - \frac{1}{7} + \frac{1}{9} - \cdots, \tag{1}$$

also known as *Leibnitz's series*, and the series $\sum_n 1/n^2$, whose sum was found by Leonhard Euler in 1734:

$$\frac{\pi^2}{6} = 1 + \frac{1}{2^2} + \frac{1}{3^2} + \frac{1}{4^2} + \frac{1}{5^2} + \cdots. \tag{2}$$

The first series is usually known as the *Gregory–Leibnitz* series but recent findings have shown that it was known to mathematicians in Kerala well before Gregory and Leibnitz or the invention of the calculus. Evidently the discoveries were made independent of one another.

Preliminaries on Infinite Series

Infinite series are fun to play with—often very pretty identities emerge from their study. Sometimes they are easy to evaluate. We have, for instance, $1/2 + 1/4 + 1/8 + 1/16 + \cdots = 1$, and more generally,

$$x + x^2 + x^3 + x^4 + \cdots = \frac{x}{1-x} \quad \text{for } |x| < 1. \tag{3}$$

Geometric series of this type are well known at the high-school level, and we shall not say more about them.[1] Another class of series that are easy to evaluate are the ones that yield to partial fraction decomposition followed by telescoping. For instance, the series $1/2 + 1/6 + 1/12 + 1/20 + 1/30 + 1/42 + \cdots$ is summed as shown below:

$$\begin{aligned}\frac{1}{2}+\frac{1}{6}+\frac{1}{12}+\cdots &= \sum_{n=1}^{\infty}\frac{1}{n(n+1)}\\ &= \sum_{n=1}^{\infty}\left(\frac{1}{n}-\frac{1}{n+1}\right)\\ &= \left(\frac{1}{1}-\frac{1}{2}\right)+\left(\frac{1}{2}-\frac{1}{3}\right)+\left(\frac{1}{3}-\frac{1}{4}\right)+\cdots\\ &= 1. \end{aligned} \tag{4}$$

after a wholesale cancellation of fractions. More challenging is the evaluation of the series $\sum_n 1/n(n+1)(n+2)$. A partial fraction decomposition yields:

$$\frac{1}{n(n+1)(n+2)} = \frac{1}{2(n+2)} - \frac{1}{n+1} + \frac{1}{2n} \tag{5}$$

and a rearrangement of terms gives

$$\sum_{n=1}^{\infty}\frac{1}{n(n+1)(n+2)} = \frac{1}{4}. \tag{6}$$

(COMMENT: We must mention here that rearranging the terms of an infinite convergent series in order to compute its sum is not always a valid step. For instance, the series $1 - 1/2 + 1/3 - 1/4 + \cdots$ converges, yielding a sum of $\ln 2$, but by rearranging the terms in different ways we get different sums. For instance, when rearranged as shown below,

$$1+\frac{1}{3}-\frac{1}{2}+\frac{1}{5}+\frac{1}{7}-\frac{1}{4}+\frac{1}{9}+\frac{1}{11}-\frac{1}{6}+\cdots,$$

with two positive terms followed by one negative term, the series converges to $\ln 2\sqrt{2}$, and the series

$$1+\frac{1}{3}+\frac{1}{5}+\frac{1}{7}-\frac{1}{2}+\frac{1}{9}+\frac{1}{11}+\frac{1}{13}-\frac{1}{4}+\cdots,$$

[1] The well known paradox of Achilles and the tortoise is resolved by studying the underlying geometric series.

with three positive terms followed by one negative term, converges to $\ln 2\sqrt{3}$. The notion of *absolute convergence* helps to make sense of these facts. Suppose the the series $\sum_i a_i$ converges. Then the series converges 'absolutely' if $\sum_i |a_i|$ converges. A theorem of Riemann's (extremely surprising at first encounter) states that if $\sum_i a_i$ converges but not absolutely (that is, $\sum_i |a_i|$ diverges), then by suitably rearranging the terms we can get the resulting series to converge to any desired number whatever! If the a_i are all of one sign, then the notions of convergence and absolute convergence coincide. In the example at hand, rearrangement can be avoided by observing that

A theorem of Riemann's (extremely surprising at first encounter) states that if $\sum_i a_i$ converges but not absolutely (that is, $\sum_i |a_i|$ diverges), then by suitably rearranging the terms we can get the resulting series to converge to any desired number whatever!

$$\frac{1}{n(n+1)(n+2)} = \frac{1}{2}\left(\frac{1}{n(n+1)} - \frac{1}{(n+1)(n+2)}\right).$$

Use of the telescoping principle now yields (6).)

Another pretty result is provided by the summation

$$x + 2x^2 + 3x^3 + 4x^4 + \cdots = \frac{x}{(1-x)^2}, \tag{7}$$

which holds whenever $|x| < 1$.

However a very different proposition is the alternating series

$$1 - \frac{1}{3} + \frac{1}{5} - \frac{1}{7} + \frac{1}{9} - \cdots. \tag{8}$$

It is not hard to show that the series converges, for by combining pairs of terms it can be written as

$$2\left(\frac{1}{3} + \frac{1}{15} + \frac{1}{35} + \frac{1}{99} + \cdots\right) = 2\sum_{n=1}^{\infty} \frac{1}{4n^2-1}. \tag{9}$$

Since $4n^2 - 1 > n(n-1)$ for $n \geq 1$, it follows that

$$\sum_{n\geq 1} \frac{1}{4n^2-1} < \frac{1}{3} + \sum_{n\geq 2} \frac{1}{n(n-1)} = \frac{4}{3}.$$

Thus the series has a well-defined sum. What is the sum? The 'school-boy' evaluation (very likely the method used by Gregory) proceeds thus: since

$$1 - x^2 + x^4 - x^6 + x^8 - \cdots = \frac{1}{1+x^2} \text{ for } |x| < 1, \tag{10}$$

we deduce, by integration between 0 and $t(|t| < 1)$, that

$$\tan^{-1} t = t - \frac{t^3}{3} + \frac{t^5}{5} - \frac{t^7}{7} + \cdots. \tag{11}$$

Substituting the value $t = 1$ in (11), we obtain the required sum as $\pi/4$. However the question of convergence is left open in this summation (an infinite series cannot always be integrated term-by-term in this fashion), and we need to fill in the necessary details. This is accomplished as follows. We consider the algebraic identity (note that this is *not* an infinite sum)

$$\frac{1}{1+x^2} = 1 - x^2 + x^4 - \cdots + (-1)^n x^{2n} + (-1)^{n+1}\frac{x^{2n+2}}{1+x^2}. \tag{12}$$

Integrating both sides between 0 and 1, we obtain

$$\frac{\pi}{4} = 1 - \frac{1}{3} + \frac{1}{5} + \cdots + (-1)^n \frac{1}{2n+1} + (-1)^{n+1}\int_0^1 \frac{x^{2n+2}}{1+x^2}dx. \tag{13}$$

The integral on the right side tends to 0 as $n \to \infty$, because

$$\left|\int_0^1 \frac{x^{2n+2}}{1+x^2}dx\right| < \int_0^1 x^{2n+2}dx = \frac{1}{2n+3}, \tag{14}$$

which tends to 0 as $n \to \infty$. The result follows.

Another approach is via evaluation of the integral I_n defined by

$$I_n = \int_0^{\pi/4} \tan^n x dx \tag{15}$$

for positive integers n. (This integral is routinely encountered while studying integration by parts). We claim that $I_n \to 0$ as $n \to \infty$. For proof we argue as follows. Obviously $I_n > 0$. Next, the tangent function is convex over the open interval $(0, \pi/4)$, so the chord connecting the points $(0,0)$ and $(\pi/4, 1)$ lies entirely above the tangent curve. The equation of this line is $y = (4/\pi)x$, so it follows that $\tan x < (4/\pi)x$ for all x in the open interval $(0, \pi/4)$. Therefore

$$0 < I_n = \int_0^{\pi/4} \tan^n x dx < \left(\frac{4}{\pi}\right)^n \int_0^{\pi/4} x^n dx$$
$$= \left(\frac{4}{\pi}\right)^n \left(\frac{\pi}{4}\right)^{n+1} \frac{1}{n+1}$$
$$= \frac{\pi}{4(n+1)}, \qquad (16)$$

and the result follows (let $n \to \infty$). Next, integration by parts yields

$$I_n = \frac{1}{n-1} - I_{n-2}. \qquad (17)$$

From (17) and the fact that $I_0 = \pi/4$, we deduce that

$$1 - \frac{1}{3} + \frac{1}{5} - \cdots + (-1)^n \frac{1}{2n-1} = \frac{\pi}{4} + (-1)^n I_{2n}. \qquad (18)$$

Letting $n \to \infty$, we obtain the Gregory–Nīlakaṇṭha series.

This approach yields an unexpected bonus:
Since $I_1 = \ln \sec \pi/4 = (1/2) \ln 2$, we obtain

$$\ln 2 = 1 - \frac{1}{2} + \frac{1}{3} - \frac{1}{4} + \frac{1}{5} - \cdots. \qquad (19)$$

(EXERCISE: Prove (19).)

The Gregory–Nīlakaṇṭha Series

Both proofs presented above make use of calculus, and it would seem difficult to arrive at the stated result without using integration. However the Kerala school of mathematics that flourished in the fifteenth and sixteenth centuries seems to have been in possession of just such a proof, due originally to the mathematician Nīlakaṇṭha[2] (1444–1550). We now present this elegant proof. In retrospect it has a certain naturalness to it, for π is after all connected to the circumference of a circle. Though calculus is not used at any stage in the proof, the intuitive notion of limit is clearly present in the author's mind.

[2] Or so we think! The facts are unfortunately not too well known. In the text Yuktibhāsā, the series is credited to Mādhava of Sangamagrāma (1350–1410), who lived almost a full century before Nīlakaṇṭha. The result is stated as follows: if c is the circumference of a circle of diameter d, then $c = 4d - \frac{4d}{3} + \frac{4d}{5} - \frac{4d}{7} + \frac{4d}{9} - \cdots$. See S Balachandra Rao in Suggested Reading for further details.

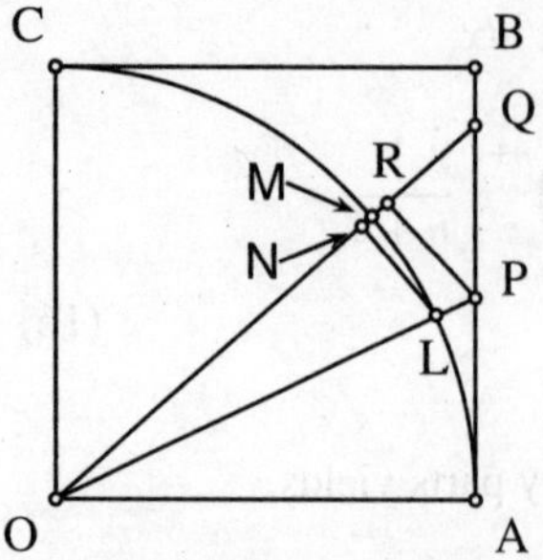

Figure 9.1

In Figure 9.1 $OABC$ is a square of side 1, and AC is a circular arc of unit radius. The length of the arc AC is thus $\pi/2$. Let P, Q be two points on the side AB, the points occurring in the order shown (A, P, Q, B) with P, Q very close to one another, i.e., with $PQ << 1$. Let OP and OQ meet the arc at L and M, respectively, and let PR and LN be the perpendiculars from P and L upon line OQ. Then $\Delta OLN \sim \Delta OPR$ and $\Delta QRP \sim \Delta QAO$. Therefore

$$\frac{LN}{OL} = \frac{PR}{OP}, \quad \text{and} \quad \frac{PR}{PQ} = \frac{OA}{OQ}. \tag{20}$$

These equations give, since $OL = OA = 1$,

$$LN = \frac{PQ}{OP \times OQ}. \tag{21}$$

Now $OP \times OQ \approx OQ^2 = 1 + AQ^2$, and arc $LM \approx LN$. Therefore

$$\text{arc } LM \approx \frac{PQ}{1 + AQ^2}. \tag{22}$$

Now let the side AB be divided into n segments of length $1/n$ each, where n is large, and let the points of subdivision be $P_1, P_2, \ldots, P_{n-1}$; let $P_0 = A, P_n = B$. Then $P_i P_{i+1} = 1/n$ for $i = 0, 1, 2, \ldots, n-1$. Summing n relations of the type (22) we obtain:

$$\frac{\pi}{4} \approx \sum_{i=1}^{n} \frac{1/n}{1 + i^2/n^2}. \tag{23}$$

(EXERCISE: Fill in the details of the proof of this formula.)

It follows that

$$\frac{\pi}{4} = \lim_{n\to\infty} \sum_{i=1}^{n} \frac{1/n}{1 + i^2/n^2}. \tag{24}$$

(The discerning reader will note at this point that we have arrived in an unexpected fashion at the 'discretized' version of the integral $\int_0^1 dx/(1 + x^2)$!) The limit is evaluated as follows. We write the fraction $(1/n)/(1 + i^2/n^2)$ as the sum of an infinite series, using the fact that $1/(1 + x^2) = 1 - x^2 + x^4 - x^6 + \cdots$ for $|x| < 1$:

$$\frac{1/n}{1 + i^2/n^2} = \frac{1}{n}\left(1 - \frac{i^2}{n^2} + \frac{i^4}{n^4} - \frac{i^6}{n^6} + \cdots\right). \tag{25}$$

Therefore

$$\frac{\pi}{4} = \lim_{n\to\infty} \frac{1}{n}\left(\sum_{i=1}^{n} 1 - \sum_{i=1}^{n} \frac{i^2}{n^2} + \sum_{i=1}^{n} \frac{i^4}{n^4} - \sum_{i=1}^{n} \frac{i^6}{n^6} + \cdots\right)$$

$$= \lim_{n\to\infty}\left(1 - \frac{\sum i^2}{n^3} + \frac{\sum i^4}{n^5} - \frac{\sum i^6}{n^7} + \cdots\right), \tag{26}$$

all summations in (26) from $i = 1$ to $i = n$.

Nīlakaṇṭha observed that the limits of the individual terms in this expression were easy to evaluate. The modern evaluation of $\lim_{n\to\infty}\left(\sum i^2\right)/n^3$ proceeds thus: since $\sum_{i=1}^{n} i^2 = n(n+1)(2n+1)/6$, the required limit is

$$\lim_{n\to\infty} \frac{n(n+1)(2n+1)}{6n^3} = \frac{1}{3} + \lim_{n\to\infty}\left(\frac{1}{2n} + \frac{1}{6n^2}\right) = \frac{1}{3}. \tag{27}$$

It is clear that Nīlakaṇṭha was aware of this as well as each of the following limits (though the language of limits was non-existent at that time):

$$\lim_{n\to\infty} \frac{\sum_{i=1}^{n} i^3}{n^4} = \frac{1}{4}, \quad \lim_{n\to\infty} \frac{\sum_{i=1}^{n} i^4}{n^5} = \frac{1}{5}, \tag{28}$$

and more generally, for any positive integer k,

$$\lim_{n\to\infty} \frac{\sum_{i=1}^{n} i^k}{n^{k+1}} = \frac{1}{k+1}. \tag{29}$$

The Nīlakaṇṭha–Gregory series,

$$\frac{\pi}{4} = 1 - \frac{1}{3} + \frac{1}{5} - \frac{1}{7} + \cdots, \tag{30}$$

now follows immediately.

If we study the proofs of Gregory and Nīlakaṇṭha carefully, we see that they both depend upon the expansion of $1/(1+x^2)$ as an infinite series.

Euler's Summation of $\sum 1/n^2$

Another series that arises naturally in many settings is the series $\sum_{n=1}^{\infty} 1/n^2$. It is easy to show that the series converges; for $n^2 > n(n-1)$, so $1/n^2 < 1/(n-1) - 1/n$. Therefore for any positive integer $k > 1$,

$$\sum_{n=1}^{k} \frac{1}{n^2} = 1 + \sum_{n=2}^{k} \frac{1}{n^2} < 1 + \sum_{n=2}^{k}\left(\frac{1}{n-1} - \frac{1}{n}\right)$$

$$= 1 + \left(1 - \frac{1}{k}\right) < 2, \tag{31}$$

What is the value of $\sum_n 1/n^2$? The question remained a mystery and Jakob Bernoulli expressed the feelings of his contemporaries when he wrote: "...If anyone finds and communicates to us that which till now has eluded our efforts, great will be our gratitude...". In 1734 Euler produced a solution using a marvellous piece of reasoning.

that is, the sum $\sum_{n=1}^{k} 1/n^2$ is less than 2. Since the partial sums of $\sum_n 1/n^2$ form a monotone increasing sequence and are bounded above by 2, they must possess a limit. This was quite clear to the contemporaries of Newton, Leibnitz and the Bernoulli brothers. But just what is the value of this limit? A numerical evaluation is easy to do and the sum can be obtained to any desired degree of accuracy if one is willing to do some computation, but can the series be evaluated in 'closed form'? The question remained a mystery and Jakob Bernoulli expressed the feelings of his contemporaries when he wrote: "... If anyone finds and communicates to us that which till now has eluded our efforts, great will be our gratitude ...". In 1734 Euler produced a solution using a marvellous piece of reasoning. The solution is a gem, and produces a truly unexpected result, but it has some questionable aspects to it, and would fall short of today's standards of rigour.

Euler started with the power series for $\sin x$:

$$\sin x = x - \frac{x^3}{3!} + \frac{x^5}{5!} - \frac{x^7}{7!} + \cdots. \tag{32}$$

This result was of course available to Leibnitz and his contemporaries; we refer to it today as the *Maclaurin series* for $\sin x$. From this it follows that for $x > 0$,

$$\frac{\sin \sqrt{x}}{\sqrt{x}} = 1 - \frac{x}{3!} + \frac{x^2}{5!} - \frac{x^3}{7!} + \cdots. \tag{33}$$

Next, Euler made the following observation: if $f(x)$ is a polynomial with roots $\alpha, \beta, \gamma, \cdots$, such that $f(0) = 1$, then

$$\frac{1}{\alpha} + \frac{1}{\beta} + \frac{1}{\gamma} + \cdots = -f'(0). \tag{34}$$

The observation is easy to justify, for the hypotheses tell us that

$$f(x) = \left(1 - \frac{x}{\alpha}\right)\left(1 - \frac{x}{\beta}\right)\left(1 - \frac{x}{\gamma}\right)\cdots. \tag{35}$$

Expanding the right side we obtain

$$f(x) = 1 - \left(\frac{1}{\alpha} + \frac{1}{\beta} + \frac{1}{\gamma} + \cdots\right) x + \cdots \tag{36}$$

and the stated claim follows. (Actually Euler did not write $f'(0)$ on the right side of (34); rather he claimed that $\sum 1/\alpha = -1\times$ the coefficient of x in $f(x)$. Since the coefficient of x in any polynomial $f(x)$ is equal to $f'(0)$, the two statements are equivalent to one another.) Euler's claim can be stated in another form: Let $f(x)$ be a polynomial with roots $\alpha, \beta, \gamma, \ldots$, none of which is 0. Then

$$\frac{1}{\alpha} + \frac{1}{\beta} + \frac{1}{\gamma} + \cdots = -\frac{f'(0)}{f(0)}. \tag{37}$$

In this form the proof is still easier: the hypothesis tells us that

$$f(x) = k(x-\alpha)(x-\beta)(x-\gamma)\ldots, \tag{38}$$

for some constant k. Taking logarithms on both sides, we obtain

$$\ln f(x) = \ln k + \ln(x-\alpha) + \ln(x-\beta) + \ln(x-\gamma) + \cdots. \tag{39}$$

Differentiating both sides of (39) we obtain

$$\frac{f'(x)}{f(x)} = \frac{1}{x-\alpha} + \frac{1}{x-\beta} + \frac{1}{x-\gamma} + \cdots, \tag{40}$$

and the substitution $x = 0$ in (40) yields (37).

Euler then took a rather large leap: *he considered the function* $f(x) = \sin\left(\sqrt{x}\right)/\sqrt{x}$ *to be a 'polynomial of infinite degree'*, namely that given by (33). This 'polynomial' certainly satisfies the relation $f(0) = 1$, because

$$\lim_{x\to 0^+} \frac{\sin\sqrt{x}}{\sqrt{x}} = 1. \tag{41}$$

(Euler did not use the terminology of limits—the notion of limit was relatively undeveloped in his time, and later Cauchy set the notion on firm ground.)

Now what are the roots of the 'polynomial' $f(x)$? Since the roots of $\sin x$ are $n\pi$, where n takes all possible integral values, the roots of $\sin\sqrt{x}$ are $n^2\pi^2$, with n taking all possible integral values. Therefore the roots of the equation $f(x) = 0$ are just the values $n^2\pi^2 (n \neq 0)$.

Finally, what is $f'(0)$ for this infinite degree polynomial? Since we already have the Maclaurin expansion of $f(x)$, namely

$$f(x) = 1 - \frac{x}{6} + \frac{x^2}{120} - \frac{x^3}{5040} + \cdots, \tag{42}$$

we see that $f'(0) = -1/6$.

Combining all these deductions into one, we see that

$$\sum_{n=1}^{\infty} \frac{1}{n^2\pi^2} = \frac{1}{6}, \tag{43}$$

and so

$$\sum_{n=1}^{\infty} \frac{1}{n^2} = \frac{\pi^2}{6}. \tag{44}$$

The required sum has been obtained—its value is $\pi^2/6$. What a miraculous achievement!

Euler's method yields other such results. Following the comments made earlier, we proceed to write $f(x)$ as an infinite product:

$$f(x) = \left(1 - \frac{x}{\pi^2}\right)\left(1 - \frac{x}{4\pi^2}\right)\left(1 - \frac{x}{9\pi^2}\right)\cdots, \tag{45}$$

that is,

$$\frac{\sin\sqrt{x}}{\sqrt{x}} = \prod_{n=1}^{\infty}\left(1 - \frac{x}{n^2\pi^2}\right). \tag{46}$$

Now the coefficient of x^2 in the product

$$\left(1 - \frac{x}{\alpha}\right)\left(1 - \frac{x}{\beta}\right)\left(1 - \frac{x}{\gamma}\right)\cdots \tag{47}$$

is $\sum 1/\alpha\beta$, which can be written as

$$\frac{1}{2}\left(\left(\sum \frac{1}{\alpha}\right)^2 - \left(\sum \frac{1}{\alpha^2}\right)\right), \tag{48}$$

the summation being over the roots of $f(x) = 0$. Since the roots are $n^2\pi^2 (n \neq 0)$, we see that

$$\left(\sum \frac{1}{\alpha}\right)^2 = \left(\sum_{n=1}^{\infty} \frac{1}{n^2\pi^2}\right)^2 = \frac{1}{6^2} = \frac{1}{36}, \tag{49}$$

and

$$\left(\sum \frac{1}{\alpha^2}\right) = \left(\frac{1}{\pi^4}\right)\left(\sum_{n=1}^{\infty} \frac{1}{n^4}\right). \tag{50}$$

Finally, the coefficient of x^2 in $f(x)$ (remember that $f(x)$ is still being thought of as a polynomial!) is $f''(0)/2$. To evaluate $f''(0)$ we use L'Hopital's rule. Differentiating the function f given by $f(x) = \sin\sqrt{x}/\sqrt{x}$ twice in succession, we find that

$$f''(x) = \frac{1}{4x^2}\left(\frac{(3 - x)\sin\sqrt{x}}{\sqrt{x}} - 3\cos\sqrt{x}\right). \tag{51}$$

Letting x tend to 0, we obtain $f''(0) = 1/60$ (left as an exercise for the reader!). Therefore

$$\frac{1}{2}\left(\frac{1}{36} - \frac{1}{\pi^4}\sum_{n=1}^{\infty}\frac{1}{n^4}\right) = \frac{f''(0)}{2} = \frac{1}{120}, \tag{52}$$

and this yields the beautiful result

$$\sum_{n=1}^{\infty} \frac{1}{n^4} = \frac{\pi^4}{90}. \tag{53}$$

A few additional remarks can be made here. Euler could equally have used the function $\cos\sqrt{x}$ instead of $\sin\sqrt{x}/\sqrt{x}$. For, writing $f(x) = \cos\sqrt{x}$, we find that the roots of $f(x) = 0$ are $(2n-1)^2\pi^2/4$ where n takes all possible positive integral values. Since $f(0) = 1$, this gives us the representation of $f(x)$ as an infinite product:

$$\cos\sqrt{x} = \prod_{n=1}^{\infty}\left(1 - \frac{4x}{(2n-1)^2\pi^2}\right). \tag{54}$$

On the other hand, $\cos x = 1 - x^2/2! + x^4/4! - x^6/6! + \cdots$, so for $x > 0$,

$$\cos\sqrt{x} = 1 - \frac{x}{2} + \frac{x^2}{24} - \frac{x^3}{720} + \cdots. \tag{55}$$

Arguing as we did earlier (the reader is invited to fill in the missing steps), we find that

$$\sum_{n=1}^{\infty} \frac{1}{(2n-1)^2} = \frac{\pi^2}{8}. \tag{56}$$

Finally, it is trivial to show that

$$\sum_{n=1}^{\infty} \frac{1}{(2n-1)^2} = \left(1 - \frac{1}{2^2}\right)\left(\sum_{n=1}^{\infty} \frac{1}{n^2}\right), \tag{57}$$

and so the result that $\sum_n 1/n^2 = \pi^2/6$ follows.

Rather remarkably, Euler's method can be used to find the sum of the Gregory–Nīlakaṇṭha series as well. (This did not escape Euler's attention—but then few things did!) We consider the function $f(x) = 1 - \sin x$, for which $f(0) = 1$, $f'(0) = -1$, $f'(0)/f(0) = -1$. The roots of the equation $f(x) = 0$ are $x = 2n\pi + \pi/2 = (4n+1)\pi/2$, where n takes all possible integral values. Moreover, these roots are all *double* roots, as a glance at the associated graph will show. (The x-axis is tangent to the graph of $y = 1 - \sin x$ at these values.) We therefore obtain, using Euler's rule:

$$-1 = -2\left(\sum_{n\in\mathbf{Z}} \frac{2}{(4n+1)\pi}\right). \tag{58}$$

(Here $\mathbf{Z}$ is the set of integers, i.e., $\{0, \pm 1, \pm 2, \pm 3, \ldots\}$.) Writing the right side in expanded form and multiplying both sides by $\pi/4$, we obtain the series of Gregory and Nīlakaṇṭha!

Euler used such methods to evaluate the value of the zeta function $\zeta(k)$ for many positive even integers k. (The zeta function is defined by

$$\zeta(t) = 1 + \frac{1}{2^t} + \frac{1}{3^t} + \cdots = \sum_{n=1}^{\infty} \frac{1}{n^t} \tag{59}$$

where t is allowed to take complex values as well.) It is known now that $\zeta(k)$ is a rational multiple of π^k whenever k is a positive even integer. Interestingly, comparatively little is known about $\zeta(k)$ when k is odd, not even whether the values taken are irrational! (It *is* known that $\zeta(3)$ is irrational—but this is an extremely recent discovery.)

As a tailpiece, we leave the following assertion for the reader to ponder over and to exercise some Eulerian ingenuity. Let the real positive roots of the equation $\tan x = x$ be $\lambda_1, \lambda_2, \lambda_3, \ldots$ Then

$$\sum_{i=1}^{\infty} \frac{1}{\lambda_i^2} = \frac{1}{10}. \tag{60}$$

(Warning!: Unlike the sine and cosine functions, the tangent function takes infinite values. How does one get around this?)

Concluding Remarks

What are we to make of the various proofs presented above? In general one can say that there are two kinds of proofs. There are proofs that are complete in every respect, with every minor detail attended to; Euclid's proofs of the infinitude of the primes and of the irrationality of $\sqrt{2}$ (both of which are to be found in *The Elements*) are certainly of this variety. Then there are proofs with a few steps missing which can however be inferred from the main body of the proof. *In particular there is no such thing as an unrigorous proof.* An unrigorous proof is no proof at all!—it is only a suggestion of a proof, and considerable work needs to be done before it can be referred to as a proof. How do the above presentations measure up in the light of these remarks?

It is clear that the proofs of the Gregory–Nīlakaṇṭha series are complete in every respect, while Nīlakaṇṭha's proof is only slightly incomplete. The missing steps are very minor and can

It is clear that the proofs of the Gregory–Nīlakaṇṭha series are complete in every respect, while Nīlakaṇṭha's proof is only slightly incomplete. The missing steps are very minor and can easily be filled in, without effort. Considering the era in which the proof was written—a time when the notion of limit was non-existent elsewhere in the world—this is

easily be filled in, without much effort. Considering the era in which the proof was written—a time when the notion of limit was non-existent elsewhere in the world—this is a remarkable achievement on the author's part. (*En passant,* one wonders what happened to the very remarkable school of thinkers to which Mādhava and Nīlakaṇṭha belonged: in what directions did they branch out? What kinds of questions did they occupy themselves with?)

a remarkable much achievement on the author's part.

However when we come to Euler's evaluation of $\sum_n 1/n^2$, we have quite a different situation before us. While we can sit back and marvel at the sheer virtuosity and brilliance of Euler's work, at its freshness and vitality, we are at the same time forced to admit that it is far from being a proof. To what extent can we apply a result to a power series when all we know is that it holds for polynomials?—that is, when can we express a function x as a product of linear expressions in x, using only the roots of $f(x)$ to go by? The answer is: not always. Euler was certainly lucky that his audacious attempt worked; as they say, first time lucky! In general, it is *not* true that if f is a complex-valued function with no singularities and with roots $\alpha_1, \alpha_2, \alpha_3, \ldots$, none of which is 0, then

However when we come to Euler's evaluation of $\sum_n 1/n^2$, we have quite a different situation before us. While we can sit back and marvel at the sheer virtuosity and brilliance of Euler's work, at its freshness and vitality, we are at the same time forced to admit that it is far from being a proof.

$$f(x) = f(0)\left(1 - \frac{x}{\alpha_1}\right)\left(1 - \frac{x}{\alpha_2}\right)\left(1 - \frac{x}{\alpha_3}\right)\cdots. \tag{61}$$

Here is a simple counterexample. Consider the function $f(x)$ defined on the complex plane thus: $f(0) = 1$, and $f(x) = (e^x - 1)/x$ for $x \neq 0$. Then f is continuous and differentiable over the whole plane, and its roots are the values $2n\pi i$, where i is $\sqrt{-1}$ and n takes all possible non-zero integral values. We may now be led to infer (after combining pairs of factors) that

$$\frac{e^x - 1}{x} = \prod_{n=1}^{\infty}\left(1 + \frac{x^2}{4n^2\pi^2}\right), \tag{62}$$

but this is absurd, because the function represented by the expression on the right side is an even function of x, whereas the function on the left side is not. (In the correct representation, an additional factor, namely $e^{x/2}$, has to be included on the right side of (62).)

In general there are infinitely many different complex-valued functions that share the same set of roots (with regard to position as well as order); thus, the functions $f(x), e^x f(x), e^{x^2} f(x), \ldots$, all share the same sets of roots. (Here x is allowed to take all possible complex values.) This holds because the exponential function e^x never takes the value 0. For a complete statement of the relevant theorems, the reader should refer to the books by Knopp, Titchmarsh and Conway in Suggested Reading or any equivalent text in complex analysis.

By the end of the nineteenth century, complex analysis had been placed on an extremely secure foundation. Obviously, this owes in considerable measure to the pioneering work done by Euler.

So Euler certainly was lucky; but then Euler had an uncanny intuition about such matters, and he probably knew just when it was safe to reason in this manner, just as Ramanujan did more than a century later. However there is still more to be said. Probably such audacity to some extent provided the impetus for the vast amount of research done in complex analysis in the century following Euler—by Cauchy, Weierstrass, Abel, Gauss, Riemann, . . .: the most powerful analysts known to mathematics. As a result, by the end of the nineteenth century, complex analysis had been placed on an extremely secure foundation. Obviously, this owes in considerable measure to the pioneering work done by Euler.

Suggested Reading

CN Srinivasiengar. *The History of Ancient Indian Mathematics*. World Press. Calcutta, 1967

T A Sarasvati Amma. *Geometry in Ancient and Medieval India*. Motilal Banārasidāss. Delhi, 1979

John B Conway. *Functions of One Complex Variable*. Springer Verlag. 2nd Edn., 1980.

Ranjan Roy. The Discovery of the Series Formula for π by Leibniz, Gregory and Nīlakaṇṭha. *Mathematics Magazine*. published by the MAA. Vol.63. No.5, December 1990.

S Balachandra Rao. *Indian Mathematics and Astronomy–Some Landmarks*. Jnana Deep Publications. Bangalore, 1994

Konrad Knopp. *Theory of Functions* Part II. Dover Books

John Stillwell. *Mathematics and its History*. Springer Verlag. Undergraduate Texts in Mathematics.

E C Titchmarsh. *A Theory of Functions*. Oxford University Press.

Shailesh A Shirali
Rishi Valley School
Rishi Valley 517 352
Andhra Pradesh.

Addendum to 'Nīlakaṇṭha, Euler and π' (*Resonance*, May 1997)

A letter has been received from M D Srinivas of the Centre for Policy Studies, Chennai. He points out that it is incorrect to attribute the result that $\pi/4 = 1 - 1/3 + 1/5 - 1/7 + \cdots$ to Nīlakaṇṭha Somayāji as was done in the article, and that it would be more appropriate to call it the *Mādhava–Gregory series*. He writes: "…[proofs] are to be found in the commentaries on the Tantrasangraha and Līlāvatī by Saṅkara Variyar (1500–1560), a student of Nīlakaṇṭha,… Saṅkara clearly attributes these series to Mādhava (1350–1410) and the proofs are presumably due to him. In any case there is no mention of Nīlakaṇṭha as having been the first to have proved them.… Some scholars (notably R C Gupta and G C Joseph) have preferred to call it the Mādhava–Gregory series…". He adds that Nīlakaṇṭha's works display the in-depth knowledge that he had of the work of Mādhava (for instance, he cites the result of Mādhava giving the value of π correct to 11 decimal places, and also explains that π cannot be exactly expressed as a ratio of two integers) and that his more important contributions lie in the field of astronomy rather than in mathematical analysis. Srinivas writes: "Nīlakaṇṭha's main contribution is that he carried out a major revision of the traditional Indian planetary model, which he presented first in his Tantrasangraha (written in 1500 AD)". Readers who are interested in learning more about this are referred to the article in *Current Science*, Volume 66, page 784 (1994), by K Ramasubramanian, M D Srinivas and M S Sriram. They could also write directly to M D Srinivas (Centre for Policy Studies, 27 Rajasekaran Street, Mylapore, Chennai 600 004).

Shailesh A Shirali

Notes

1. With reference to equation (4), it should be noted that telescoping ("wholesale cancellation of fractions") works only because the terms in the brackets tend to 0. That is, the sum

$$u_1 + \sum_{n\geq 1}(u_{n+1} - u_n)$$

converges to u_1 only if $u_n \to 0$ as $n \to \infty$. For example, telescoping will not work for the sum $\sum_{n=1}^{\infty}\left(\sqrt{n+1} - \sqrt{n}\right)$, which obviously diverges.

2. With reference to equation (60), we must mention that there are a great many mysteries connected with the zeta function $\zeta(t)$ defined for $t \in \mathbf{C}$ with $\Re(t) > 1$ by

$$\zeta(t) = 1 + \frac{1}{2^t} + \frac{1}{3^t} + \cdots.$$

The series converges only when $\Re(t) > 1$, but $\zeta(t)$ may be defined as a meromorphic function on $\mathbf{C}\setminus\{1\}$, with a simple pole at $t = 1$, using a device called analytic continuation. Indeed we have $\zeta(0) = -1/2, \zeta(-1) = -1/12, \zeta(-3) = 1/120, \ldots$ As noted in the text, a lot is known about $\zeta(t)$ when t is an even positive integer; for instance we have Euler's result:

$$\zeta(2n) = (-1)^{n+1}\frac{(2\pi)^{2n} \cdot B_{2n}}{2 \cdot (2n)!} \quad (n \in \mathbf{N}). \tag{1}$$

The B's here are the Bernoulli numbers, defined via the generating function $x/(e^x - 1)$:

$$\frac{x}{e^x - 1} = B_0 + B_1 x + B_2 x^2 + B_3 x^3 + \cdots;$$

thus $B_0 = 1, B_1 = -1/2, B_2 = 1/6, \ldots$. It happens that $x/(e^x - 1) + x/2$ is an even function of x, so $B_{2n+1} = 0$ for $n \in \mathbf{N}$. Formula (1) implies that $\zeta(2n)$ is a rational multiple of π^{2n} for all $n \in \mathbf{N}$. It is curious however that we know little about $\zeta(n)$ for odd $n \in \mathbf{N}$. For instance, we know that $\zeta(3)$ is irrational, but not whether it is a rational multiple of π^3.

In addition to these unanswered questions about the zeta function we also have the Riemann hypothesis—perhaps *the* most famous and most important unsolved mathematical problem today. This is the statement that the only non-trivial zeroes of the zeta function occur when $\Re(t) = 1/2$. (The 'trivial' zeroes occur at the negative even integers, $t = -2, -4, -6, \ldots$. They may be explained by the formula

$$\zeta(-n) = \frac{(-1)^n B_{n+1}}{n+1} (n \in \mathbf{N}).$$

Since $B_{2n+1} = 0$ for $n \in \mathbf{N}$, the statement that $\zeta(-2n) = 0$ for $n \in \mathbf{N}$ follows.) There is an enormous quantity of numerical evidence in support of the Riemann hypothesis (RH for short), but no proof is currently in sight. For interest's sake we mention that the first two zeroes of $\zeta(t)$ on this line occur approximately at $t = 0.5 + 14.13473i$ and $t = 0.5 + 21.02204i$.

Concerning why the RH is considered so important, we mention only the fact that it may be stated in an equivalent form in terms of the function $\pi(x)$ giving the number of primes less than or equal to x. The connection between the zeta function and this function was first described by Riemann in a seminal and highly influential paper. The interested reader should refer to the book *Riemann's Zeta Function* by H M Edwards (Academic Press).

10

Stirling's Formula

S Ramasubramanian

We will derive an estimate for $n!$ The approach taken here is as in Feller's book (W. Feller. *An Introduction to Probability Theory and its Applications*, Vol. 1, Wiley-Eastern, 1993). For this we estimate $\log(n!) = \log 1 + \log 2 + \cdots + \log n$.

As log is a monotone function note that for $k \geq 1$

$$\int_{k-1}^{k} \log x\, dx < \log k < \int_{k}^{k+1} \log x\, dx.$$

Summing the above over $1 \leq k \leq n$ we get

$$\int_{0}^{n} \log x\, dx < \log(n!) < \int_{1}^{n+1} \log x\, dx.$$

Since $\frac{d}{dx}(x \log x) = 1 + \log x$, we now have

$$(n \log n) - n < \log(n!) < \big((n+1)\log(n+1)\big) - n. \qquad (1)$$

The two-sided inequality (1) suggests comparing $\log(n!)$ with the arithmetic mean of the extreme members of (1) or with some quantity 'close' to the arithmetic mean.

Note that

$$\frac{(n \log n) + (n+1)\log(n+1)}{2} - \left(n + \frac{1}{2}\right)\log n \to \frac{1}{2}.$$

So we consider comparing $\log(n!)$ with $\left(n + \frac{1}{2}\right)\log(n) - n$; (why is it a reasonable thing to do?).

Put

$$d_n = \log(n!) - \left(n + \frac{1}{2}\right)\log(n) + n. \tag{2}$$

Using the expansion $\log\left(\frac{1+t}{1-t}\right) = 2\sum_{j=0}^{\infty} \frac{1}{(2j+1)} t^{(2j+1)}$, for $|t| < 1$, we now get

$$\begin{aligned}
d_n - d_{n+1} &= \left(n + \frac{1}{2}\right)\log\left(\frac{n+1}{n}\right) - 1\\
&= \left(n + \frac{1}{2}\right)\log\left(\frac{1+\frac{1}{2n+1}}{1-\frac{1}{2n+1}}\right) - 1\\
&= \frac{1}{3(2n+1)^2} + \frac{1}{5(2n+1)^4} + \cdots\\
&< \frac{1}{3}\left[\frac{1}{(2n+1)^2} + \frac{1}{(2n+1)^4} + \cdots\right]\\
&= \frac{1}{3}\frac{1}{((2n+1)^2 - 1)} = \frac{1}{12n(n+1)}\\
&= \frac{1}{12n} - \frac{1}{12(n+1)}.
\end{aligned} \tag{3}$$

By the first three equalities in (3) we get $d_n > d_{n+1}$, whereas from the inequality of the extreme terms of (3) we get $d_n - \frac{1}{12n} < d_{n+1} - \frac{1}{12(n+1)}$. These imply that the sequence $\{d_n\}$ converges to a finite limit. (Why?)

Put $\log C_0 = \lim_n d_n$. Then by (2) we have $n!/C_0 e^{-n} n^{n+\frac{1}{2}} \to 1$ as $n \to \infty$. Thus we have proved that there is a constant $C_0 > 0$ such that

$$\lim_{n\to\infty} \frac{n!}{C_0 e^{-n} n^{n+\frac{1}{2}}} = 1. \tag{4}$$

It turns out (as a consequence of De Moivre–Laplace limit theorem), that $C_0 = \sqrt{2\pi}$; (4) is known as *Stirling's formula*.

S Ramasubramanian
Statistics and Mathematics Unit
Indian Statistical Institute
RVCE Post
Bangalore 560 059
Karnataka

11

Fourier Series

The Mathematics of Periodic Phenomena

S Thangavelu

There are several natural phenomena that are described by periodic functions. The position of a planet in its orbit around the sun is a periodic function of time; in chemistry, the arrangement of molecules in crystals exhibits a periodic structure. The theory of Fourier series deals with periodic functions. By a periodic function we mean a function f of a real variable t which satisfies the relation $f(t+c) = f(t)$ for all t, where c is a constant. Here c is called the *period* of f. The simplest examples of periodic functions of period 2π are provided by the trigonometric functions $\sin kt$ and $\cos kt$ for integer k. It follows that linear combinations of the form

$$P_n(t) = \sum_{k=0}^{n} (a_k \cos kt + b_k \sin kt)$$

are also periodic with the same period 2π. These are called *trigonometric polynomials*. The basic idea behind Fourier series is that any periodic function (of period 2π) can be approximated by trigonometric polynomials. (Notice that it is enough to consider periodic functions with period 2π because a function f with some other period, say c, can be converted to a function of period 2π by a suitable change of variable. (Exercise!)) The subject of Fourier series finds a wide range of applications from crystallography to spectroscopy. It is one of the most powerful theories in the history of mathematics and has stimulated the development of several branches of analysis.

Consider the function $f(t)$ defined by the formula

$$f(t) = \begin{cases} \frac{\pi}{2} - t, & 0 \le t \le \pi \\ \frac{\pi}{2} + t, & -\pi \le t \le 0, \end{cases}$$

which for other values of t is extended by periodicity. Thus the graph[1] of the function looks as shown in Figure 11.1.

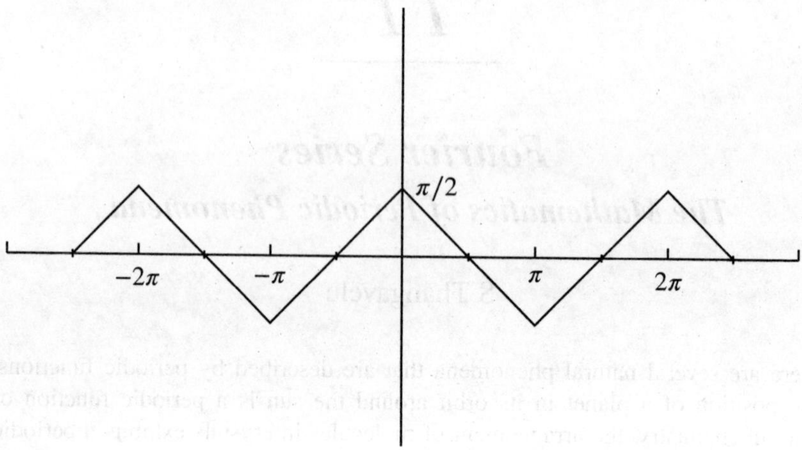

Figure 11.1

This function is continuous but not differentiable at the points $0, \pm\pi, \pm 2\pi, \cdots$ where the graph has corners. The function $f(t)$ is even. The functions $\cos kt$ are all even and the functions $\sin kt$ are all odd.[2] Hence we may expect to approximate f by trigonometric polynomials of the form $\sum_{k=0}^{l} a_k \cos kt$. In fact consider

$$P_n(t) = \frac{4}{\pi} \sum_{k=0}^{n} (2k+n)^{-2} \cos(2k+1)t.$$

The graphs of these functions for $n = 2$ and for $n = 4$ are shown in Figure 11.2.

From the picture it is clear that even for small values of n, the trigonometric polynomial P_n approximates f very well.

Now you may be wondering why we have chosen the particular trigonometric polynomials P_n in order to approximate f. As we have already remarked the absence of the sine terms may be explained by the fact that f is even and $\sin kt$ are all odd; but why did we choose the coefficients $(2k+n)^{-2}$ for the terms $\cos(2k+1)t$? Is there a way of relating the function f to the coefficients $(2k+n)^{-2}$? To answer all these questions, let us go back in history by a couple of centuries. The basic idea that any periodic function may be expanded in terms of trigonometric polynomials is attributed to D Bernoulli. This Swiss mathematician believed in the 'universal character' of the trigonometric polynomials much before Fourier. It all started with his works in 1747, 1748 and 1753 where he studied the vibrating

[1] The author thanks A S Vasudevamurthy for helping him with the various graphs in this article. The material in some of the boxes and footnotes were provided by C. Varughese.

[2] A function h is *even* if $h(x) = h(-x)$ for all x and *odd* if $h(x) = -h(-x)$ for all x.

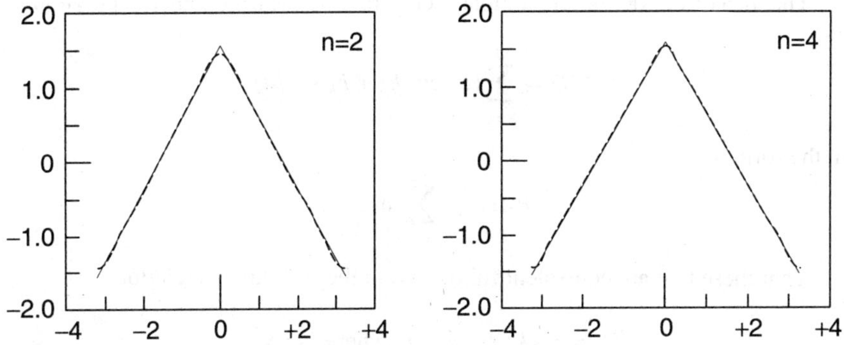

Figure 11.2

string problem: find the solutions of the following partial differential equation with the given initial conditions

$$\frac{\partial^2 u(x,t)}{\partial t^2} = \frac{\partial^2 u(x,t)}{\partial x^2}$$

$$u(x,0) = f(x), u(-\pi,t) = u(\pi,t) = 0$$

This is called the *wave equation* and the graph of $u(x,t)$ represents the shape of the vibrating string at time t. The initial shape of the string is given by the function f and the condition $u(-\pi,t) = 0 = u(\pi,t)$ means that the string is kept fixed at the ends.

This equation had been studied by Euler and D' Alembert before Bernoulli. D'Alembert gave the solution of the above problem in the form

$$u(x,t) = \frac{1}{2}[f(x+t) + f(x-t)].$$

We can easily check that this function u is indeed a solution provided f is an odd periodic function with period 2π and is twice differentiable. But Bernoulli had a different idea. The functions

$$u_k(x,t) = a_k \cos kt \sin kx$$

satisfy the wave equation, and following physical ideas, Bernoulli suggested solutions of the form

$$u(x,t) = \sum_{k=0}^{\infty} a_k \cos kt \sin kx.$$

Based on this observation Bernoulli was led to believe in the possibility of expanding arbitrary periodic f with $f(\pi) = f(-\pi) = 0$ in terms of $\sin kx$ but he did not have a clue as to how to calculate the various coefficients!

Mathematics had to wait for almost fifty years for a formula to calculate the *Fourier coefficients*. In 1807, the French mathematician Jean Joseph Fourier working on the heat equation found a way to calculate the coefficients.

The idea of Fourier is very simple. Let us rewrite the trigonometric polynomials

$$P_n(t) = \sum_{k=0}^{n} (a_k \cos kt + b_k \sin kt)$$

in the form

$$P_n(t) = \sum_{k=-n}^{n} \alpha_k e^{ikt}.$$

That these two are equivalent follows from the well known relation

$$e^{ikt} = \cos kt + i \sin kt, \text{ where } i = \sqrt{-1}.$$

Suppose that these trigonometric polynomials $P_n(t)$ converge to $f(t)$ in any reasonable sense so that we can take the limit under the integral sign to get

$$\lim_{n\to\infty} \int_{-\pi}^{\pi} P_n(t)e^{-imt}\,dt = \int_{-\pi}^{\pi} f(t)e^{-imt}\,dt, \text{ for each fixed } m.$$

The integrals on the left hand side of the above equation can be easily calculated. In fact

$$\int_{-\pi}^{\pi} P_n(t)e^{-imt}\,dt = \sum_{k=-n}^{n} \alpha_k \int_{-\pi}^{\pi} e^{i(k-m)t}\,dt$$

and an easy integration reveals that

$$\int_{-\pi}^{\pi} e^{i(k-m)t}\,dt = 0$$

whenever $k \neq m$. And when $k = m$ we get 2π for the value of the integral. Thus we have the formula for α_m:

$$\alpha_m = \frac{1}{2\pi} \int_{-\pi}^{\pi} f(t)e^{-imt}\,dt.$$

To emphasize that α_m depends on f we write $\alpha_m = \hat{f}(m)$. The associated series

$$f(t) = \sum_{k=-\infty}^{\infty} \hat{f}(k)e^{ikt}$$

is called the *Fourier series* of the function f.[3] $\left\{\hat{f}(k)\right\}_{k=-\infty}^{\infty}$ are called the *Fourier coefficients*.

---

[3] The equality here must be taken to mean that the partial sums of the series on the right approximate the function f in a sense that will be described later. The two sides are not necessarily equal for every value of t, although it is customary to write it in this form.

EXERCISE: If f is differentiable and f' is continuous, show that the Fourier series corresponding to f' is given by

$$\sum_{k=-\infty}^{\infty} (ik)\hat{f}(k)e^{ikt}.$$

Fourier's idea is so simple that one may wonder why it eluded D'Alembert, Bernoulli, and even the great Euler. But one has to bear in mind that these mathematicians of the seventeenth century had to deal with concepts like functions, integrals and the convergence of infinite series when they were not well understood. For Euler, a function always meant an analytic expression; Bernoulli believed that any possible position of a vibrating string represents a function and can be given by an analytic expression. It was Fourier who went one step further to consider all functions as either continuous or discontinuous, and boldly asserted that they can be represented by an infinite series which came to be known as Fourier series. The moral of the story is: a simple argument can still be very deep and it may not be very simple to discover!

Let us go back and calculate the Fourier coefficients of the function f which we started with. As f is even

$$\hat{f}(0) = \frac{1}{2\pi}\int_{-\pi}^{\pi} f(t)dt$$

$$= \frac{1}{\pi}\int_{0}^{\pi}\left(\frac{\pi}{2} - t\right) dt = 0.$$

For $k \neq 0$,

$$\hat{f}(k) = \frac{1}{2\pi}\int_{-\pi}^{\pi} f(t)e^{-ikt}dt$$

$$= \frac{1}{2\pi}\int_{-\pi}^{\pi} f(t)\cos ktdt,$$

since the integral

$$\int_{-\pi}^{\pi} f(t)\sin ktdt = 0$$

on account of the fact that $\sin kt$ is odd. Therefore,

$$\hat{f}(k) = \frac{1}{\pi}\int_{0}^{\pi}\left(\frac{\pi}{2} - t\right)\cos ktdt$$

$$= -\frac{1}{\pi}\int_{0}^{\pi} t\cos ktdt,$$

since the integral $\int_0^\pi \cos kt dt = 0$. Integrating by parts the last integral we can easily see that

$$\hat{f}(k) = \frac{1}{\pi k^2}(1 - \cos k\pi) = \frac{1}{\pi k^2}(1 - (-1)^k).$$

The last equality holds because $\cos k\pi = (-1)^k$. Thus $\hat{f}(2k) = 0$ and $\hat{f}(2k+1) = \frac{2}{\pi}(2k+1)^{-2}$, and consequently the Fourier series of f becomes

$$f(t) = \frac{2}{\pi} \sum_{k=-\infty}^{\infty} (2k+1)^{-2} e^{i(2k+1)t}.$$

Combining the terms corresponding to $(2k+1)$ and $-(2k+1)$ we get

$$f(t) = \frac{4}{\pi} \sum_{k=0}^{\infty} (2k+1)^{-2} \cos(2k+1)t.$$

Now, it should be clear why the polynomials $P_n(t)$ gave good approximations to the function $f(t)$!

Thanks to Fourier, we now have a way of writing down the Fourier series of any periodic function. But what is the guarantee that the Fourier series actually represents the function? In other words, how do we know if the partial sums defined by

$$S_n f(t) = \sum_{k=-n}^{n} \hat{f}(k) e^{ikt}$$

converge to the function f pointwise, i.e., how do we know that $S_n f(t)$ converges to $f(t)$ as $n \to \infty$, for every t? It is not very hard to prove that for the particular function of Figure 11.1, we have such a convergence. Figure 11.2 is a convincing evidence of this fact. Fourier believed and asserted that, in general, $S_n f$ always converges to the function f, but he offered no proof. *Fourier's assertion turned out to be wrong as we will see shortly*. This by no means diminishes the greatness of Fourier's contributions. In fact, with a different interpretation of convergence, his assertion is correct: the Fourier series of a function converges in what the modern day mathematicians call the *sense of distributions*.

The theory of distributions—an important development of this century—was actually anticipated by Fourier as one can see from his treatment of Fourier integrals! For another mode of convergence, which is valid for a very wide class of functions, see Box 1.

The pointwise convergence of the partial sums $S_n f$ to the function f fails—it fails dramatically at some points as we see from the following example. Consider the function $g(t) = -1$ for $-\pi \leq t \leq 0, g(t) = 1$ for $0 < t \leq \pi$. This function is

Box 1

If we do not insist on pointwise convergence, but are willing to settle for *mean square convergence*, then indeed for a wide class of 2π–periodic functions f, the Fourier series converges to f. All we need to assume is that f is *square-integrable* i.e. $\int_{-\pi}^{\pi} \left| f(x) \right|^2 dx$ is finite. For such $f, S_n f$ converges to f in the mean-square sense even if for *many* values of t, $S_n f(t)$ does *not* converge to $f(t)$ i.e. $\frac{1}{2\pi}\int_{-\pi}^{\pi} \left| S_n f(t) - f(t) \right|^2 dt \to 0$ as $n \to \infty$ i.e. the *average value* of the square of the error $\left| S_n f(t) - f(t) \right| \to 0$ as $n \to \infty$. If f represents a periodic signal, physicists and electrical engineers will tell you that $\int_{-\pi}^{\pi} \left| f(t) \right|^2 dt$ represents the *energy* in the signal. Using the fact about mean square convergence explained above, we can actually prove that $\int_{-\pi}^{\pi} \left| f(t) \right|^2 dt = 2\pi \sum_{k=-\infty}^{\infty} \left| \hat{f}(k) \right|^2$. However instead of using $\exp(ikt)$, if we use $\sin kt$ and $\cos kt$ and write the Fourier series as $a_0 + \sum_{k=1}^{\infty} b_k \cos kt + \sum_{k=1}^{\infty} c_k \sin kt$, then the above formula can be written as $\int_{-\pi}^{\pi} \left| f(t) \right|^2 dt = 2\pi \left| a_0 \right|^2 + \pi \sum_{k=1}^{\infty} \left| b_k \right|^2 + \pi \sum_{k=1}^{\infty} \left| c_k \right|^2$ (why?). Finally if g is another 2π–periodic function with corresponding Fourier series $A_0 + \sum_{k=1}^{\infty} B_k \cos kt + \sum_{k=1}^{\infty} C_k \sin kt$, then one has the more general formula $\int_{-\pi}^{\pi} f(t)g(t)dt = 2\pi a_0 A_0 + \pi \sum_{k=1}^{\infty} b_k B_k + \pi \sum_{k=1}^{\infty} c_k C_k$.

discontinuous at the points $0, \pm\pi, \pm 2\pi, \ldots$, where it has a jump of size 2. We can easily calculate the Fourier series of this function. We leave it as an exercise to the reader to show that $\hat{g}(0) = 0, \hat{g}(2k) = 0$ and that $\hat{g}(2k+1) = -i\frac{2}{\pi}(2k+1)^{-1}$. Therefore, the Fourier series of g takes the form

$$g(t) = \frac{-2i}{\pi} \sum_{k=-\infty}^{\infty} (2k+1)^{-1} e^{(2k+1)it},$$

or combining the terms with $(2k+1)$ and $-(2k+1)$ we can write it as

$$g(t) = \frac{4}{\pi} \sum_{k=0}^{\infty} (2k+1)^{-1} \sin(2k+1)t.$$

Consider now the partial sums associated to the above Fourier series:

$$S_n g(t) = \frac{4}{\pi} \sum_{k=0}^{n} (2k+1)^{-1} \sin(2k+1)t.$$

The graphs of some of these trigonometric polynomials are shown in Figure 11.3.

From the graphs we observe that $S_n g$ approximates g well except in a small neighbourhood of $t = 0$ where it overshoots and undershoots the levels $+1$ and -1 respectively. Naturally, one expects that the overshoot and undershoot tends

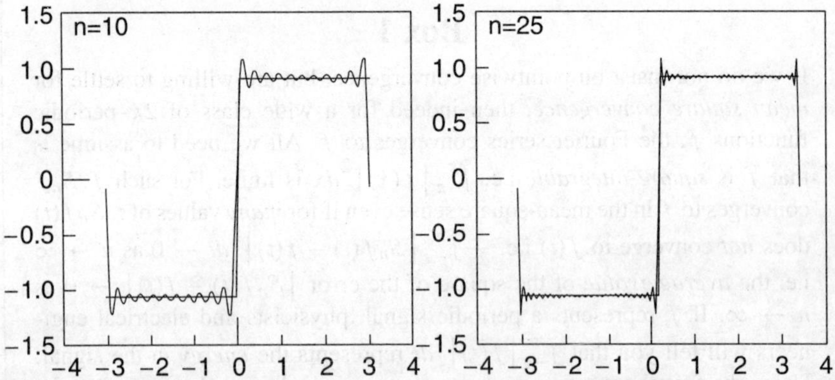

Figure 11.3

to zero as n goes to infinity; but surprisingly this does not take place. Instead, they are always about 9%. This is called the *Gibbs phenomenon* because it was pointed out by Gibbs in a letter to *Nature* in 1899. The letter was a reply to one from Michelson (of the *Michelson–Morley experiment* fame), who apparently was angry with the machine he used for computing Fourier series, as it failed badly at jumps!

The Gibbs phenomenon that occurs in the vicinity of jump discontinuities can be proved rigorously. This already shows that we cannot expect the Fourier series to converge at all points. One may think that the situation will be different in the case of continuous functions. But already, in 1876, du Bois-Reymond constructed a continuous function whose Fourier series fails to converge at a given point. Several eminent mathematicians including Dirichlet, Riemann and Cantor occupied themselves with the problem of convergence of Fourier series. Many positive results were proved in the course of time, but in 1926 the Russian mathematician A N Kolmogorov came up with a discouraging result. He constructed an integrable (in the sense of Lebesgue) function whose Fourier series diverges everywhere! After this example by Kolmogorov, mathematicians started believing that one day someone will construct a continuous function with everywhere-divergent Fourier series. That day never came! Instead, in 1966, the Swedish mathematician L Carleson showed that in the case of continuous functions the Fourier series can fail to converge only on a set which is *negligible* in a certain sense (i.e., of *measure* zero)!

When the function f is smooth, say once differentiable, and f' is continuous, it can be shown that the Fourier series converges to the function at each point, and even *uniformly*. But in applications we have to deal with functions having very bad discontinuities. Dirichlet and Dini found local conditions on the functions that will ensure pointwise convergence of the Fourier series. These results are adequate for the purpose of applications. However, mathematicians not being content with such results have been seeking more and more general results, proving the convergence of Fourier series for larger and larger classes of functions! They have also worked with different notions of convergence, although unfortunately, many of these cannot be explained here without assuming a lot of Lebesgue theory of integration.

Box 2

Let $f(t) = t, -\pi \leq t < \pi$, and extended for other values of t by periodicity. Use the result at the end of Box 1 to calculate the value of π^2 as the sum of an infinite series.

We would like to conclude this article with the following result of Fejer which treats the class of continuous functions as a whole. As we know, given any point t_0, there is a function in this class whose Fourier series diverges at that point. In 1904, the Hungarian mathematician Fejer had the brilliant idea of considering the *averages of partial sums instead of the partial sums themselves*. Thus he considered

$$\sigma_n f = \frac{1}{n+1}(S_0 f + S_1 f + \cdots S_n f).$$

These are called the *Cesaro means* and it is easy to see that they can be written in the form

$$\sigma_n f(t) = \sum_{k=-n}^{n} \left(1 - \frac{|k|}{n+1}\right) \hat{f}(k) e^{ikt}.$$

The celebrated theorem of Fejer says that when f is continuous, the Cesaro means $\sigma_n f$ converge to f, not only pointwise, but also in the stronger sense of *uniform convergence.* Hence trigonometric polynomials are *dense* in the class of continuous functions!

Box 3

Notice that the functions $\sin kt$ and $\cos kt$ are periodic with period $(2\pi/k)$. Of course they are periodic with period 2π but they are also periodic with the smaller period $(2\pi/k)$ and this is the smallest period for these functions. Thus, they have *frequency* $(k/2\pi)$. By choosing various values of k, we get what we can think of as a *standard family* of signals. If f is a periodic signal (with period 2π), its Fourier series is a decomposition of f into signals from the standard family. The relation $\int_{-\pi}^{\pi} |f(t)|^2 dt = 2\pi \sum |\hat{f}(k)|^2$ (see Box 1) is then a description of how the energy contained in the signal f is distributed among various frequencies. The relative amount of energy contained at any frequency is determined by the square of the modulus of the corresponding Fourier coefficient.

The body of literature dealing with Fourier series has reached epic proportions over the last two centuries. We have only given readers an outline of the topic in this article. For the full length episode we refer the reader to the monumental treatise of A Zygmund. Beginners will find the books of Bhatia and Körner very helpful. For more advanced readers, the books of Dym–McKean, Folland, Helson and Katznelson make enjoyable reading. The article of Gonzalez–Velasco deals with the influence of Fourier series on the development of analysis and we highly recommend it.

Suggested Reading

A Zygmund. *Trigonometric Series.* Volumes I and II. Cambridge University Press. 1959.

H Dym and H P McKean. *Fourier Series and Integrals*. Academic Press, 1972.

Y Katznelson. *An Introduction to Harmonic Analysis*. Dover Publications, 1976.

T W Körner. *Fourier Analysis*. Cambridge University Press, 1989.

G B Folland. *Fourier Analysis and its Applications*. Wadsworth and Brooks, 1992.

E Gonzalez–Velasco. *Connections in Mathematical Analysis, the Case of Fourier Series*. American Mathematical Monthly. pp 427–441, May 1992.

R Bhatia. *Fourier Series*. TRIM-2. Hindusthan Book Agency, 1993.

H Helson. *Harmonic Analysis* (2nd edition). TRIM-7. Hindusthan Book Agency, 1995.

S Thangavelu
Indian Statistical Institute
Bangalore 560 059
Karnataka

12

The Isoperimetric Problem

A Sitaram

"Among all simple closed curves[1] γ on the plane having a fixed perimeter, the circle encloses the maximum area". This is the celebrated isoperimetric problem and one of the proofs (due to Hurwitz) uses results about Fourier series for this purely geometrical statement! [An equivalent statement is: "Among all simple closed curves γ on the plane enclosing a given area, the circle has the least perimeter". (Exercise: Why are the two statements equivalent?)] For the necessary background on Fourier series, the reader can consult the article by S Thangavelu in this volume.

In fact, the statement above is a consequence of the following inequality:

$$L^2 - 4\pi A \geq 0, \tag{1}$$

where L is the *perimeter* of the simple closed curve and A is the area enclosed, and *equality is attained in the above if and only if γ is a circle*. (Exercise: Convince yourself that this implies the assertion made in the beginning of the first paragraph.)

We make certain simplifying assumptions that are not really very serious. Let us assume the curve γ is traced out by a moving point p whose coordinates are $(\gamma_1(t), \gamma_2(t))$. Let us also assume that at time $t = 0$, we are at position p_0 and that we are back at the same position at time $t = 2\pi$ (see Figure 12.1).

By repeating this over and over again, we can think of $\gamma_1(t)$ and $\gamma_2(t)$ as periodic functions of t with period 2π. To simplify matters, we take $\gamma_1(t)$ and $\gamma_2(t)$ to be *smooth* functions, in particular $\gamma_1'(t)$ and $\gamma_2'(t)$ exist and are continuous functions. (The case of *non-smooth* curves can be taken care of by an appropriate

[1] By a simple closed curve on the plane, we mean the path traced by a point p on the plane, which is initially at p_0 at time $t = a$, and returns to p_0 at a later time $t = b$ *without* crossing its path at any other time between a and b. To avoid pathologies, we assume that all our curves are *rectifiable*, i.e., have *finite length*!. This is called the *perimeter* of the curve.

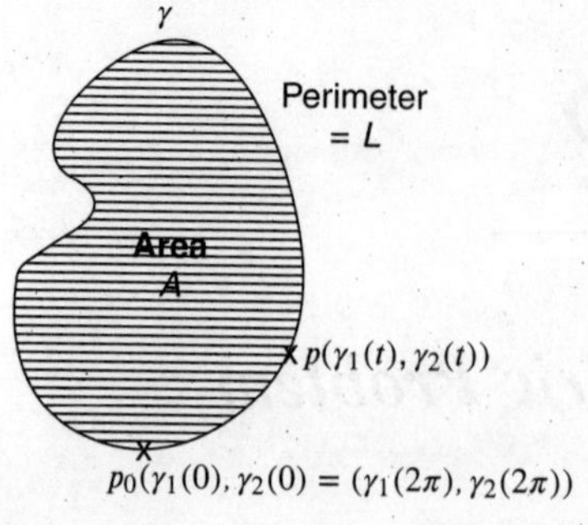

Figure 12.1

approximation procedure.) Since the perimeter L and the area A enclosed do not depend on how exactly we move around the curve, let us move in an *anti-clockwise manner with constant speed u.* Thus $\left(\gamma_1'(t)\right)^2 + \left(\gamma_2'(t)\right)^2 = u^2$, for all t. But, since the total length is L and each cycle is completed in 2π units of time, $u = \frac{L}{2\pi}$, and so we have $\frac{L^2}{4\pi^2} = \left(\gamma_1'(t)\right)^2 + \left(\gamma_2'(t)\right)^2$ for all t.

Thus,

$$\int_0^{2\pi} \frac{L^2}{4\pi^2} dt = \frac{L^2}{2\pi} = \int_0^{2\pi} [(\gamma_1'(t))^2 + (\gamma_2'(t))^2]dt. \tag{2}$$

Since, as has already been pointed out, $\gamma_1(t)$ and $\gamma_2(t)$ are 2π-periodic functions of t, we can express them in the form of *Fourier series:*

$$\gamma_1(t) = a_0 + \sum_{n=1}^{\infty} a_n \cos nt + \sum_{n=1}^{\infty} b_n \sin nt, \tag{3A}$$

$$\gamma_2(t) = A_0 + \sum_{n=1}^{\infty} A_n \cos nt + \sum_{n=1}^{\infty} B_n \sin nt. \tag{3B}$$

So,

$$\gamma_1'(t) = \sum_{n=1}^{\infty} nb_n \cos nt + \sum_{n=1}^{\infty} (-na_n) \sin nt, \tag{3C}$$

$$\gamma_2'(t) = \sum_{n=1}^{\infty} nB_n \cos nt + \sum_{n=1}^{\infty} (-nA_n) \sin nt. \tag{3D}$$

Hence, using basic facts about Fourier series,

$$\int_0^{2\pi} (\gamma_1'(t))^2 dt = \pi \sum_{n=1}^{\infty} (n^2 b_n^2 + n^2 a_n^2)$$

and

$$\int_0^{2\pi} (\gamma_2'(t))^2 dt = \pi \sum_{n=1}^{\infty} (n^2 B_n^2 + n^2 A_n^2).$$

Thus from the above and (2) we get

$$\frac{L^2}{2\pi} = \pi \sum_{n=1}^{\infty} n^2(a_n^2 + b_n^2 + A_n^2 + B_n^2). \tag{4}$$

On the other hand, from our knowledge of *advanced* calculus, we know that the enclosed area is given by the formula:

$$A = \int_0^{2\pi} \gamma_1(t)\gamma_2'(t)dt.$$

(The reader is encouraged to look up a proof of this result which relies on Green's theorem.) But again from our knowledge of Fourier series and using (3A) and (3D) we have

$$\int_0^{2\pi} \gamma_1(t)\gamma_2'(t)dt = \pi \sum_{n=1}^{\infty} n(a_nB_n - A_nb_n).$$

So,

$$A = \pi \sum_{n=1}^{\infty} n(a_nB_n - A_nb_n). \tag{5}$$

Using (4) and (5) (and some simple school algebra!), we get

$$L^2 - 4\pi A = 2\pi^2 \sum_{n=1}^{\infty} \Big[(na_n - B_n)^2 + (n^2-1)B_n^2 + (nb_n + A_n)^2 + (n^2-1)A_n^2\Big]. \tag{6}$$

The quantity on the right in (6) is clearly non-negative! Hence $L^2 - 4\pi A \geq 0$ and the proof of inequality (1) is complete.

When is equality attained in (1)? For this, the sum of the infinite series in (6) must be zero. Since each term of the infinite series is non-negative and each of the four quantities in each term is non-negative, we are led to:

$$\begin{aligned} na_n &= B_n, \quad (n^2-1)B_n^2 = 0, \\ nb_n &= -A_n, \quad (n^2-1)A_n^2 = 0, \\ &\text{for } n = 1, 2, \cdots. \end{aligned}$$

Thus $B_n = 0, A_n = 0$ for $n \geq 2$.

Also, $a_1 = B_1$ and $b_1 = -A_1$ and if we let $a_1 = B_1 = C$ and $b_1 = -A_1 = D$, from (3A) and (3B), we have

$$\gamma_1(t) = a_0 + C\cos t + D\sin t$$

$$\gamma_2(t) = A_0 - D\cos t + C\sin t$$

and so $(\gamma_1(t) - a_0)^2 + (\gamma_2(t) - A_0)^2 = C^2 + D^2$. This is clearly the equation of the circle with centre at (a_0, A_0) and radius $\sqrt{C^2 + D^2}$, and the proof of the assertion we started with is complete!

Though much simpler proofs of inequality (1) are available (see Suggested Reading, for instance the article of P D Lax), the method described in this note actually provides more information about L and A in terms of the Fourier coefficients (via equation (6)) for an *arbitrary closed curve* γ. This kind of information, linking purely *geometric data* (like L and A) with purely *analytic data* (like the Fourier coefficients), is considered very important in mathematics. For some exciting connections between *analysis* and *geometry*, we refer the reader to the beautiful (but difficult!) article by M Kac (see Suggested Reading). We conclude by pointing out that the following three dimensional analogue of the question we started with is also true (although more difficult to prove): Among all closed surfaces enclosing a given volume, the sphere has the smallest surface area. (Ask your physics teacher about the important implications of this proposition to physics!).

Suggested Reading

M Kac. Can one hear the shape of a drum? *American Mathematical Monthly*. Vol.73. pp.1–23, 1966.

T W Körner. *Fourier series*. Cambridge University Press, 1989.

P D Lax. A short path to the shortest path. *American Mathematical Monthly*. Vol.102. pp. 158–159, 1995.

S Thangavelu. Fourier series.*Resonance*. Vol.1. No.10. pp.44–55, October 1996. (Also chapter 11 in this volume)

A Sitaram
Indian Statistical Institute
R V College Post
Bangalore 560 059
Karnataka.

13

From Calculus to Wavelets—A New Mathematical Technique

Wavelet Analysis

Gerald B Folland

A 'wavelet' is a function that exhibits oscillatory behaviour in some interval and then decays rapidly to zero outside this interval. A remarkable discovery of recent years is that the translations and dilations of certain wavelets can be used to form sets of 'basic' functions for expanding general functions into infinite series, and these expansions have many theoretical and practical applications.

The expansion of functions into infinite series is one of the most powerful techniques of mathematical analysis. The general idea is to write a function $f(t)$ (defined on the real line or an interval in the real line) as

$$f(t) = \sum c_n \phi_n(t), \tag{1}$$

where the ϕ_n's are a family of 'basic' functions with properties that are significant for the problem at hand and the coefficients c_n are real or complex numbers depending on f. (As we shall see, sometimes it is more convenient to choose an index set for the ϕ's other than the natural numbers $0, 1, 2, \ldots$) For such an expansion to be useful, the coefficients c_n must be computable in terms of f in a reasonably straightforward way, and the series $\sum c_n \phi_n$ must encode important information about the function f in an easily accessible form.

Usually the first type of series expansion that one meets as a mathematics student is *power series*, that is,

$$f(t) = \sum_{n=0}^{\infty} c_n (t-a)^n \tag{2}$$

The coefficients of such a series are given by Taylor's formula,

$$c_n = \frac{f^{(n)}(a)}{n!}.$$

In particular, a function cannot be represented as a power series unless it possesses derivatives of all orders (and even this is not a guarantee of the validity of (2)). However, if f has derivatives upto order N, one can still use Taylor's formula with remainder,

$$f(t) = \sum_{n=0}^{N} \frac{f^{(n)}(a)}{n!}(t-a)^n + R_N(t), \tag{3}$$

in which the remainder R_N vanishes more rapidly than $|t-a|^N$ as $t \to a$. This formula provides a very precise description of $f(t)$ for t near a but is usually not of much use when t is far away from a.

Another very useful type of series is the *trigonometric series* or *Fourier series*. These series come in several closely related forms, of which the simplest for analytical purposes is the one involving the complex exponentials $e^{2\pi int} = \cos 2\pi nt + i \sin 2\pi nt$:

$$f(t) = \sum_{n=-\infty}^{\infty} b_n e^{2\pi int} \tag{4}$$

It was Fourier's great discovery that such series can be used to represent more or less arbitrary functions on the real line that are *periodic* with period 1 (that is, $f(t+1) = f(t)$ for all t). More precisely, if f is any periodic function—even one with lots of discontinuities and infinite singularities—for which the area under the graph of $|f|^2$ over one period is finite,

$$\int_{-1/2}^{1/2} |f(t)|^2 dt < \infty, \tag{5}$$

then f admits an expansion of the form (4). (If f is a rather 'rough' function, there may be some exceptional points where the series does not converge to $f(t)$, but it converges to $f(t)$ 'almost everywhere.') The *Fourier coefficients* b_n are given by a simple integral formula, whose meaning is explained in Box 1:

$$b_n = \int_{-1/2}^{1/2} f(t)e^{-2\pi int} dt. \tag{6}$$

One of the virtues of the Fourier representation is that it converts differentiation into a simple algebraic operation.

One of the virtues of the Fourier representation (4) is that it converts differentiation into a simple algebraic operation. Indeed, term-by-term differentiation of (4) yields

$$f'(t) = \sum_{-\infty}^{\infty} 2\pi in b_n e^{2\pi int}.$$

Box 1 The Meaning of Orthonormality

A sequence of complex-valued functions ϕ_n is said to be *orthonormal* on an interval I (which might be the whole real line) if

$$\int_I \phi_n(t)\overline{\phi_m(t)}\, dt = 0 \text{ for } n \neq m \text{ (orthogonality)}, \qquad \text{(a)}$$

$$\int_I |\phi_n(t)|^2 dt = 1 \text{ (normalization)}. \qquad \text{(b)}$$

(The terminology comes from the analogy with an orthonormal or unit-perpendicular set of vectors $\mathbf{v}_n$; the integrals $\int_I \phi_n\overline{\phi}_m$ replace the dot products $\mathbf{v}_n \cdot \mathbf{v}_m$.) For example, it is an easy exercise to check that the functions $e^{2\pi int}$ used in Fourier analysis are orthonormal on any interval of length 1. If the functions ϕ_n are orthonormal and f can be expanded in terms of them,

$$f(t) = \sum c_n\phi_n(t) \quad (t \in I), \qquad \text{(c)}$$

the coefficients c_n must be given by

$$c_m = \int_I f(t)\overline{\phi_m(t)}dt. \qquad \text{(d)}$$

To see this, multiply both sides of (c) by $\overline{\phi_m(t)}$ and integrate over I:

$$\int_I f(t)\overline{\phi_m(t)}dt = \sum_n c_n \int_I \phi_n(t)\overline{\phi_m(t)}dt.$$

By (a), the integrals on the right all vanish except the one with $n = m$, and that one equals 1 by (b), so the sum on the right reduces to c_m. Conversely, if (c) always implies (d), the sequence ϕ_n must be orthonormal. Indeed, the identity

$$\phi_m = \cdots + 0 \cdot \phi_{m-1} + 1 \cdot \phi_m + 0 \cdot \phi_{m+1} + 0 \cdot \phi_{m+2} + \cdots$$

is an instance of (c), and the corresponding instance of (d) is just (a) together with (b).

This needs some justification, but the result is correct: *Suppose f is differentiable and f' satisfies condition (5). If the Fourier coefficients of f are denoted by b_n, then the Fourier coefficients of f' are $2\pi inb_n$.* Similarly for higher derivatives: the Fourier coefficients of $f^{(k)}$ are $(2\pi in)^k b_n$.

As a consequence, it is easy to read off the order of differentiability of f from its Fourier coefficients. If f has derivatives upto order k that satisfy (5), then the series $\sum(2\pi in)^k b_n e^{2\pi int}$ must converge, and hence $n^k b_n$ must tend to zero as $|n| \to \infty$. In other words, b_n must tend to zero more rapidly than $|n|^{-k}$. Conversely, if $|b_n| \leq C|n|^{-k}$, the series (4) can be differentiated at least $k-1$ times (the $(k-1)$th derivative may exist only 'almost everywhere'). In short: *The order of smoothness of a function is reflected in the rate of decay of its Fourier coefficients.*

The bad news is that it is hard to tell from a Fourier series *where* the non-smooth behavior occurs, or what it looks like. For example, the function

$$g_1(t) = \sum_{n \neq 0} \frac{e^{2\pi int}}{2in} = \sum_{1}^{\infty} \frac{\sin 2\pi nt}{n} \tag{5}$$

has jump discontinuities at the points $0, \pm 1, \pm 2, \ldots$ but is perfectly smooth (in fact, linear) elsewhere. Its close relative

$$g_2(t) = \sum_{n \neq 0} \frac{(-1)^n e^{2\pi int}}{2in} = \sum_{1}^{\infty} \frac{(-1)^n \sin 2\pi nt}{n} \tag{6}$$

has the same sort of behavior, but with discontinuities at $\pm\frac{1}{2}, \pm\frac{3}{2}, \pm\frac{5}{2}, \ldots$. And the function fashioned from g_1 by omitting all the terms whose order is not a power of 2,

$$g_3(t) = \sum_{k=0}^{\infty} \frac{\sin 2\pi 2^k t}{2^k}, \tag{9}$$

is everywhere continuous but nowhere differentiable! See Figures 13.1-13.3

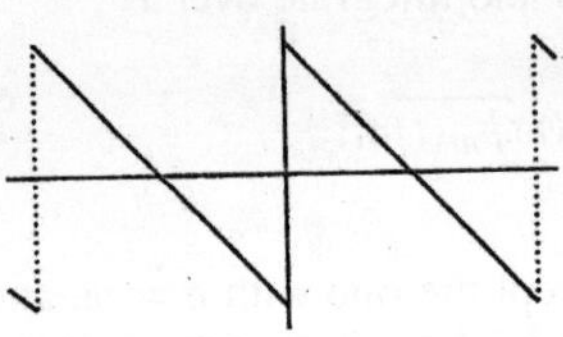

Figure 13.1 *The function g_1 defined by* (7).

We have presented Fourier series for functions that are periodic with period 1. A simple change of variable yields a corresponding expansion for functions that are periodic with period P:

$$f(t) = \frac{1}{P} \sum_{-\infty}^{\infty} b_n e^{2\pi int/P},$$

where

$$b_n = \int_{-P/2}^{P/2} f(t) e^{-2\pi int/P} \, dt.$$

In the limit as $P \to \infty$, one obtains an integral rather than a series:

$$f(t) = \int_{-\infty}^{\infty} \hat{f}(\omega) e^{2\pi i\omega t} \, d\omega \tag{10a}$$

where $\hat{f}(\omega) = \int_{-\infty}^{\infty} f(t) e^{-2\pi i\omega t} \, dt.$ (10*b*)

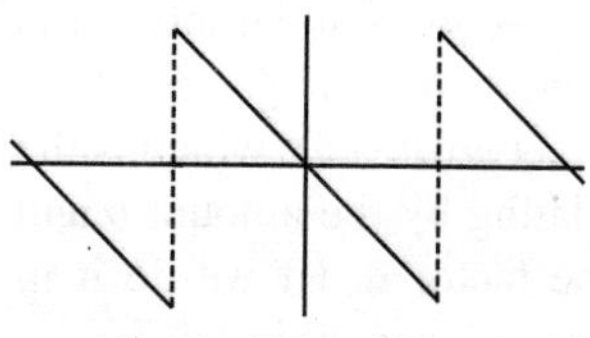

Figure 13.2 *The function* g_2 *defined by* (8).

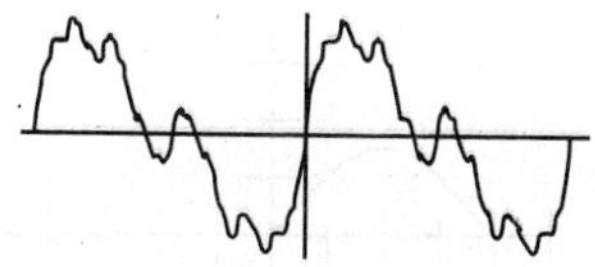

Figure 13.3 *The function* g_3 *defined by* (9).

The formula (10*a*) is called the *Fourier integral* expansion of f. It is valid for (non-periodic) functions f on the real line that satisfy the analogue of (5):

$$\int_{-\infty}^{\infty} |f(t)|^2 dt < \infty \tag{11}$$

and it expresses such functions as a continuous superposition of the simple sinusoidal functions $e^{2\pi i\omega t}$. The coefficients $\hat{f}(\omega)$ in this expansion are given by (10*b*); they constitute a function of the frequency ω called the *Fourier transform* of f. The connection between smoothness of f and the size of its Fourier coefficients, discussed above for the periodic case, continues to hold in this situation.

To summarize: Taylor series provide precise *local* information about a function, but only if the function has many derivatives. Fourier series or integrals can be used for functions with no smoothness properties, and they yield lots of information about the *global* properties of the function, but they are inefficient for analyzing the detailed behavior of a function near a point. (How could you ever tell where the discontinu ities of the functions in (7) and (8) occur by examining their Fourier coefficients?) What is missing is a method for analyzing the *local irregular* behavior of functions that aren't smooth—and this is where wavelets come into play.

Taylor series provide precise local information about a function, but only if the function has many derivatives. Fourier series or integrals can be used for functions with no smoothness properties, and they yield lots of information about the global properties of the function, but they are inefficient for analyzing the detailed behavior of a function near a point.

What is a 'wavelet'? This term is loosely used to denote a function that exhibits oscillatory behavior in some interval I, then decays rapidly to zero (or perhaps vanishes identically) outside I. If we have one such function $\psi(t)$, we can obtain others by *dilations* or *translations*—that is, by replacing the variable t by at $(a > 0)$ or $t - b(-\infty < b < \infty)$. The dilation $\psi(t) \to \psi(at)$ changes the frequency of the oscillations of ψ by a factor of a, and simultaneously changes the length of the interval on which ψ lives by a factor of $1/a$.

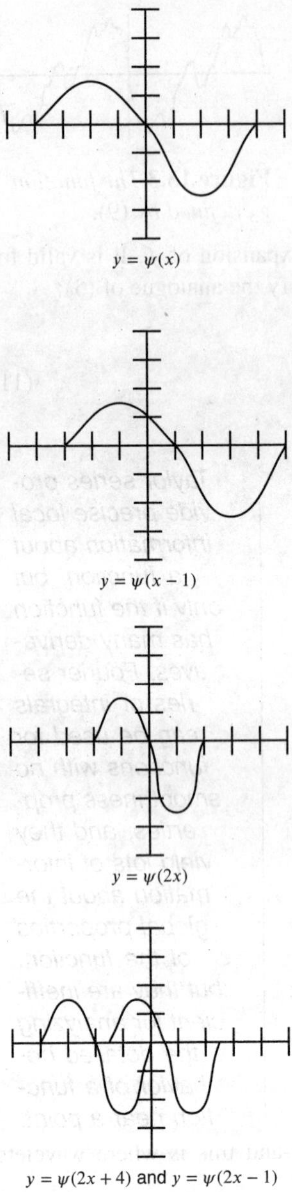

Figure 13.4 *The effect of translations and dilations on a function.*

The translation $\psi(t) \to \psi(t-b)$ simply shifts ψ over by the amount b. We can also combine these operations to get $\psi(at-b)$, which is the result of *first* translating by the amount b and *then* dilating by the factor a. (If we do it in the reverse order, we get something different: $\psi(t) \to \psi(at) \to \psi(a[t-b]) = \psi(at-ab)$.) See Figure 13.4.

Now, here is the basic problem of wavelet analysis: *Can we use the translations and dilations of a wavelet ψ to form a set of 'basic' functions for expanding general functions into infinite series as in (1)?* Let us make a formal definition: An *orthonormal wavelet* is a real- or complex-valued function $\psi(t)$ defined on the real line, with the following properties:

[1] $\psi(t)$ tends to zero faster than any power of t as $t \to \pm\infty$.

[2] ψ possesses continuous derivatives upto order N, for some positive integer N.

[3] For all integers j and n, let

$$\psi_{jn}(t) = 2^{j/2}\psi(2^j t - n). \qquad (12)$$

Then every function f on the real line that satisfies (11) can be expanded uniquely in a series of the ψ_{jn}:

$$f(t) = \sum_{j,n=-\infty}^{\infty} c_{jn}\psi_{jn}(t). \qquad (13)$$

[4] The coefficients c_{jn} in (13) are given by

$$c_{jn} = \int_{-\infty}^{\infty} f(t)\overline{\psi_{jn}(t)}dt, \qquad (14)$$

where $\overline{\psi_{jn}(t)}$ is the complex conjugate of $\psi_{jn}(t)$.

As explained in Box 1, properties [3] and [4] imply that the functions ψ_{jn} are orthonormal; hence the name orthonormal wavelet. The factor of $2^{j/2}$ in (12) is there to make $\int_{-\infty}^{\infty}|\psi_{jn}(t)|^2 dt$ independent of j and n.

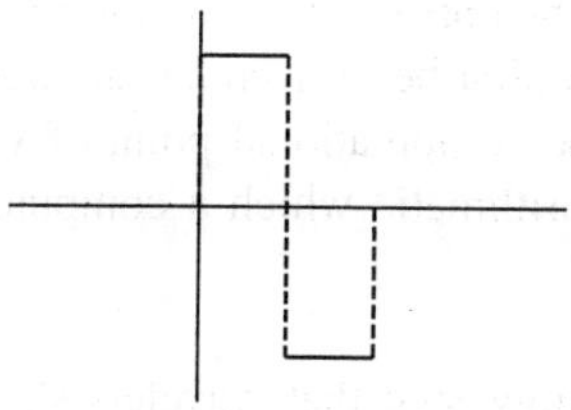

Figure 13.5 *The Haar wavelet.*

The existence of orthonormal wavelets is not at all obvious. If we omit condition [2], there is a simple example that has been known since 1910, the *Haar wavelet*

$$\psi_{\text{Haar}}(t) = \begin{cases} 1 & \text{if } 0 < t < \frac{1}{2}, \\ -1 & \text{if } \frac{1}{2} < t < 1, \\ 0 & \text{otherwise.} \end{cases}$$

See Figure 13.5. However, the Haar wavelet is too crude to be very useful. Finding orthonormal wavelets that satisfy [2] is much trickier, and it was not done until 1986–87 when a group of mathematicians led by Yves Meyer took some ideas that had been used in signal processing and other areas of applied mathematics and turned them into a rigorous mathematical theory. Meyer constructed an orthonormal wavelet that has derivatives of all orders, but its decay at infinity is relatively slow (a computational disadvantage). See Figure 13.6. Guy Battle and Pierre-Gilles Lemarié constructed a family of orthonormal wavelets, one for each order of smoothness N in [2], with exponential decay at infinity. Finally, by an ingenious adaptation of some algorithms developed by engineers for signal processing, Ingrid Daubechies constructed a family of orthonormal wavelets, with all possible finite orders of smoothness, that actually vanish outside of finite intervals. See Figure 13.7.

Finally, by an ingenious adaptation for some algorithms developed by engineers for signal processing, Ingrid Daubechies constructed a family of orthonormal wavelets, with all possible finite orders of smoothness, that actually vanish outside of finite intervals.

Let us pause to say a few more words about these constructions. Meyer's method is Fourier-analytic; he gives an explicit formula not for ψ but for its Fourier transform. The Battle–Lemarié wavelets are *splines*, that is, piecewise polynomial functions. Both these types of functions are well known in mathematical analysis. But Daubechies's wavelets are something new. They are given not by analytic formulae but as limits of recursive transformations. That is, a Daubechies wavelet ψ is obtained as the limit of a sequence of functions ϕ_k, where ϕ_0 is a simple piecewise linear function and ϕ_k is defined in terms of its predecessor ϕ_{k-1} by a recursion formula of the form

Figure 13.6 *The wavelet of Meyer type.*

$$\phi_k(t) = \sum_{n=-N}^{N} C_n \phi_{k-1}(2t - n), \tag{15}$$

and the whole secret is to find the appropriate coefficients C_n to make this process work. The resulting limit function ψ is a rather peculiar beast from the standpoint of classical mathematics, but it is lovely from the computational point of view, because the formula (15) involves only simple arithmetic which a computer is quite happy to perform over and over again!

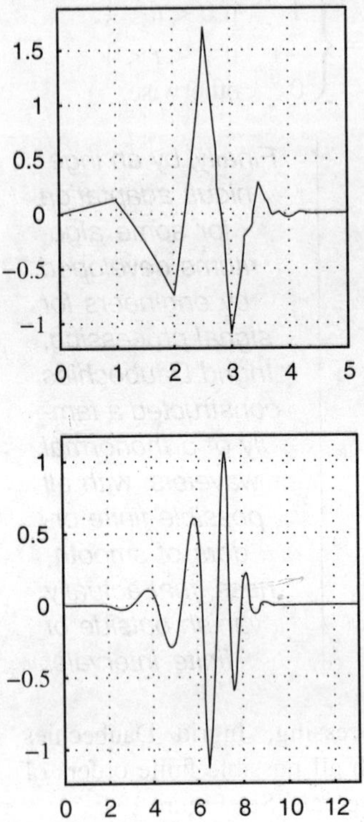

Figure 13.7 *Two of the Daubechies wavelets.*

We have suggested that wavelets should have some oscillatory behaviour, and the accompanying graphs also indicate this, but there seems to be nothing in properties [1]–[4] about oscillation. In fact, it is not too hard to show that properties [1]–[4] automatically imply the following additional property of wavelets:

[5] Let N be the integer in property [2]. Then

$$\int_{-\infty}^{\infty} \psi_{jn}(t)P(t)dt = 0 \qquad (16)$$

for all integers j and n and all polynomials P of degree $\leq N$.

This property implies that ψ must be quite oscillatory in order to produce the cancellations that make all the integrals in (16) vanish. For example, by taking $P(t) \equiv 1$ (and $j = n = 0$) we see that $\int_{-\infty}^{\infty} \psi(t)dt = 0$, which means that positive values of ψ in one region must be balanced by negative values in another.

We now examine the meaning and applications of wavelet expansions of the form (13). To simplify the discussion, we shall assume that we are working with a Daubechies wavelet ψ with N continuous derivatives that vanishes outside a bounded interval I. (Almost everything we say remains true for other wavelets if one modifies the statements to take account of the rapidly decaying tails of the wavelet.)

The function ψ_{jn} defined by (12) then vanishes outside the interval I_{jn} obtained by translating I by the amount n and then dilating by the factor 2^{-j}; thus, if I has length l then I_{jn} has length $2^{-j}l$. At the same time, we may think of the frequency of the oscillations of ψ as being on the order of magnitude of 1. (This is just a manner of speaking, as the oscillations of ψ are irregular and don't have a definite frequency.) Then the frequency of the oscillations of ψ_{jn} is on the order of magnitude of 2^j. In short, for $j << 0, \psi_{jn}$ represents *low-frequency* oscillations over *long* intervals, and for $j >> 0, \psi_{jn}$ represents *high-frequency* oscillations over *short* intervals.

We can now see how to interpret (13) and (14):

$$f = \sum_{j,n=-\infty}^{\infty} c_{jn}\psi_{jn}, \qquad c_{jn} = \int_{-\infty}^{\infty} f(t)\overline{\psi_{jn}(t)}\,dt$$

$$= \int_{I_{jn}} f(t)\overline{\psi_{jn}(t)}\,dt.$$

Let us rewrite the expansion of f as

$$f = \sum_{j=-\infty}^{\infty} S_j, \quad \text{where } S_j = \sum_{n=-\infty}^{\infty} c_{jn}\psi_{jn}. \tag{17}$$

The sum $\sum_{j<0} S_j$ represents a 'blurred' version of f that shows its large features but omits the fine details of size 1 or less. The sum S_0 adds in oscillations of frequency roughly 1 on intervals of length roughly 1; the sum S_1 adds in oscillations of frequency roughly 2 on intervals of length roughly 1/2; and so forth. Each additional sum S_j adds another level of detail at the length scale 2^{-j}.

Moreover, the individual terms in these sums are *local* in nature. The coefficient c_{jn} depends only on the values of f on the interval I_{jn}, and the term $c_{jn}\psi_{jn}$ lives on this same interval. Since these intervals only overlap to a finite extent as n varies (with j fixed), the sum S_j provides an efficient description of the local details of the behaviour of f at length scale 2^{-j}. As j increases, it is like looking at pieces of f under increasingly strong microscopes.

The smoothness of f in the neighborhood of a point $t = a$ is reflected in the decay as $j \to +\infty$ of those wavelet coefficients c_{jn} for which the interval I_{jn} contains a.

(On the other hand, when $j << 0, S_j$ might be said to represent a view of f through the wrong end of a telescope. If f is negligibly small outside an interval of length L, it is easy to see that S_j will also be negligibly small for $2^{-j} >> L$, because things on the scale of L are 'invisible' to the long-range waves ψ_{jn}. Hence the terms in (13) with $j << 0$ are not of much importance for most purposes.)

We mentioned earlier that smoothness of a periodic function is reflected in the decay of its Fourier coefficients. The same is true for wavelet expansions, but now on a local level: *The smoothness of f in the neighbourhood of a point $t = a$ is reflected in the decay as $j \to +\infty$ of those wavelet coefficients c_{jn} for which the interval I_{jn} contains a.* The reason for this fact (of which there are several much more precise versions) can be briefly explained as follows. If f has N continuous derivatives near $t = a$, we can use Taylor's formula (3) to write $f = P_N + R_N$

where P_N is the Taylor polynomial of degree N about $t = a$ and $R_N(t)$ is very small for t near a. Hence, by property [5],

$$c_{jn} = \int_{I_{jn}} f(t)\overline{\psi_{jn}(t)}dt = \int_{I_{jn}} [P_N(t) + R_N(t)]\overline{\psi_{jn}(t)}dt$$
$$= \int_{I_{jn}} R_N(t)\overline{\psi_{jn}(t)}dt.$$

But if I_{jn} contains a and j is large (so that the length of I_{jn} is small), I_{jn} will be contained in the region where R_N is very small, and hence c_{jn} will be very small!

For example, let us return to the functions g_1, g_2, and g_3 defined by (7), (8), and (9). The Fourier series for g_1 and g_2 look almost alike, but their wavelet expansions are very different because the sets of indices (j, n) for which I_{jn} contains one of the discontinuities of g_1 or g_2 are very different. It is only for those values of j and n that the coefficients c_{jn} for g_1 or g_2 will be significant for j large. (In fact, if I_{jn} does not contain a discontinuity of g_1 or g_2, C_{jn} will actually be zero by property [5], since g_1 or g_2 is linear on I_{jn}!) On the other hand, g_3 is more or less equally rough at all points, so its coefficients c_{jn} will all decay slowly as $j \to +\infty$.

Wavelets have been found to be particularly useful in problems that require separating out the parts of a signal that pertain to different scales of time or length, and in processing signals or images whose most significant features lie in the regions where rapid variations occur.

Wavelet expansions have significant applications in a number of areas, particularly in the analysis and processing of electric and acoustic signals of two-dimensional images such as photographs. (The latter require wavelets in two variables, whose theory is similar to the one-dimensional theory sketched above.) Wavelets have been found to be particularly useful in problems that require separating out the parts of a signal that pertain to different scales of time or length, and in processing signals or images whose most significant features lie in the regions where rapid variations occur. (For example, in an ordinary photograph the most important features may be edges that indicate the boundaries of the objects in the photograph; in an astronomical photograph it is important to distinguish between distant stars and even more distant galaxies, whose details show up at different length scales.) There is considerable evidence that the human brain uses something akin to wavelet analysis in processing the information it receives from the eyes and ears, and researchers in human vision were using wavelets in an informal way long before the present mathematical theory was developed.

There is considerable evidence that the human brain uses something akin to wavelet analysis in processing the information it receives from the eyes and ears.

This is not to say that wavelet expansions of the sort we have discussed are good for everything, however. For example, they are not particularly efficient for encoding musical signals because they do not provide enough frequency resolution. (Roughly speaking, the expansion (17) only provides frequency resolution to within the nearest octave, because the frequency scales specified by the index j differ by factors of 2.) However, the discovery of orthonormal wavelets has stimulated the development of a variety of related techniques for expanding functions (some of which go under the names of bi-orthogonal wavelets, wavelet packets, local Fourier bases, etc.) that offer much greater flexibility in adapting the expansion to the problem at hand. For example, one can sacrifice some precision in spatial or temporal resolution to obtain better frequency resolution, or sacrifice the perfect efficiency of the orthogonal wavelet expansion for expansions that have some built-in redundancy and hence are less sensitive to encoding errors. But this is another story.

Suggested Reading

CK Chui (ed.). *Wavelets: a Tutorial in Theory and Applications.* Academic Press. Boston, 1992. It is a collection of expository articles about wavelets and related topics.

Y Meyer. *Wavelets: Algorithms and Applications.* Society for Industrial and Applied Mathematics. Philadelphia, 1993. It covers the wavelet theory and its applications on a somewhat higher mathematical level.

R S Strichartz. How to make wavelets. *American Mathematical Monthly.* Vol.100. pp. 539–556, 1993. It is a good short exposition of the techniques for constructing waves.

J J Benedetto and M W Frazier (eds.). *Wavelets: Mathematics and Applications.* CRC Press. Boca Raton, Florida, 1994. It is a collection of expository articles about wavelets and related topics.

E Hernandez and G Weiss. *A First Course on Wavelets.* CRC Press. Boca Raton, Florida, 1996. It is a fine introductory treatise on wavelet theory.

B B Hubbard. *The World According to Wavelets.* A K Peters. Wellesley. Mass., 1996. It is a delightful non-technical account of the development of wavelet theory and its applications.

GERALD B FOLLAND
Department of Mathematics
University of Washington
Seattle, WA 98195-4350
USA

14

From Fourier Series to Fourier Transforms

A Sitaram and S Thangavelu

Thanks to the pioneering work of Bernoulli, Euler, Lagrange and Fourier, today every student of science and engineering knows that a periodic function $f(x)$ of period 2π can be expanded (possibly in an infinite series) in terms of sin kx and cos kx, $k = 0, 1, 2, \cdots$ (see [1]). For more on the historical developments that led to this idea, the reader is encouraged to consult [2]. Thus, using complex notation, a 2π-periodic function $f(x)$ on the real line $\mathbb{R}$ can be 'written' in the form

$$f(x) \approx \sum_{n=-\infty}^{\infty} a_n e^{inx}, \tag{1}$$

where, as Fourier observed, the a_n's are given by the formula $a_n = \frac{1}{2\pi}\int_0^{2\pi} f(x)e^{-inx}dx$. In what sense does the series on the right converge, and if it does converge, in what sense is it equal to $f(x)$? These are surprisingly subtle matters and the answers depend on the nature of the function $f(x)$. For instance, if, as is the case in most applications, f is square-integrable on $[0, 2\pi]$, i.e., $\int_0^{2\pi} |f(x)|^2 dx < \infty$, then the series on the right hand side of (1) is absolutely square summable i.e. $\sum_{n=-\infty}^{\infty} |a_n|^2 < \infty$, and converges in the mean square sense to f, i.e., $\int_0^{2\pi} |f(x) - \sum_{k=-n}^{n} a_k e^{ikx}|^2 dx \to 0$ as $n \to \infty$. If f is reasonably smooth, then the series converges uniformly to f. For more on this fascinating story the reader is referred to [2].

The appearance of the function e^{inx} is not entirely accidental! Note that the interval $[0, 2\pi]$, with the end points identified, can be thought of as the unit circle $S^1 = \{z \in \mathbb{C} : |z| = 1\}$, via the identification $\theta \leftrightarrow e^{i\theta}$. Under this identification, the function e^{inx} is really the function χ_n on S^1 defined by $\chi_n(z) = z^n$. S^1 is a group under multiplication and the $\chi_n' s, n = 0, \pm 1, \pm 2, \cdots$ account for *all* the continuous homomorphisms of S^1 into S^1. We now pass on to the case of functions $f(x)$ defined on $\mathbb{R}$ which satisfy a reasonable condition like

$\int_{-\infty}^{\infty}|f(x)|dx < \infty$ (integrability) or $\int_{-\infty}^{\infty}|f(x)|^2dx < \infty$ (square integrability). Notice that such functions (which are not identically zero) cannot be periodic. (Why not?) So, no matter how we re-scale the real line, we cannot really hope to 'expand' such functions in terms of $\sin kx$, and $\cos kx, k = 0, 1, 2, \cdots$, as in the periodic case. Taking the cue from the fact that the functions e^{inx} correspond to the homomorphisms χ_n of S^1 into S^1, we look for continuous homomorphisms of $\mathbb{R}$ into S^1. It turns out that *all* such continuous homomorphisms of $\mathbb{R}$ into S^1 are given by $\varphi_\lambda(x) = e^{i\lambda x}, \lambda \in \mathbb{R}$. ($\mathbb{R}$ is a group with + as the group law.) So we are tempted to conjecture that $f(x)$ can be 'expanded' in terms of $e^{i\lambda x}$. Clearly since we have a continuum of such functions, we may have to replace the infinite series in the periodic case by an integral in the non-periodic case. That this can be done is one of the great triumphs of Fourier transform theory. To cut a long story short, for an *integrable* function $f(x)$, define $\hat{f}(\lambda)$, the *Fourier transform* of $f(x)$, by $\hat{f}(\lambda) = \int_{-\infty}^{\infty} f(x)e^{-i\lambda x}dx$. Then it turns out that $\hat{f}(\lambda)$ is a continuous function on $\mathbb{R}$, and $|\hat{f}(\lambda)| \to 0$ as $|\lambda| \to \infty$. If the function f is *sufficiently well behaved*, then we can recover $f(x)$ from $\hat{f}(\lambda)$ by the Fourier inversion formula $f(x) = \frac{1}{2\pi}\int_{-\infty}^{\infty}\hat{f}(\lambda)e^{i\lambda x}d\lambda$. Moreover if f is *also square-integrable*, then we do have mean square convergence as in the Fourier series case, i.e. $\int_{-\infty}^{\infty}|f(x) - \frac{1}{2\pi}\int_{-R}^{R}\hat{f}(\lambda)e^{i\lambda x}d\lambda|^2dx \to 0$ as $R \to \infty$, and $\int_{-\infty}^{\infty}|f(x)|^2dx = \frac{1}{2\pi}\int_{-\infty}^{\infty}|\hat{f}(\lambda)|^2d\lambda$. Actually there is a purely *heuristic* way of justifying the Fourier inversion formula starting from the Fourier expansion of a periodic function. (See pages 204–205 of [3]).

There is another, perhaps intuitively more appealing way of motivating the Fourier transform. Notice that the Fourier inversion formula is an expansion of f in terms of the functions $\varphi_\lambda(x) = e^{i\lambda x}$. Now each φ_λ is an eigenfunction for the one-dimensional Laplace operator $\frac{d^2}{dx^2}$ with eigenvalue $-\lambda^2$. Among these the 2π-periodic ones are precisely $e^{ikx}, k = 0, \pm 1, \pm 2, \cdots$. (In higher dimensions, one would have to consider the usual Laplacian $\frac{\partial^2}{\partial x_1^2} + \cdots + \frac{\partial^2}{\partial x_n^2}$.) Thus the Fourier inversion (expansion) formula as well as the Fourier-series formula can be thought of as an eigenfunction expansion for the Laplacian.

Suppose f represents the amplitude of a signal (say a sound wave or a light wave) at time t. Then the formulae:

$$\hat{f}(\omega) = \int_{-\infty}^{\infty} f(t)e^{-i\omega t}dt \text{ and } f(t) = \frac{1}{2\pi}\int_{-\infty}^{\infty}\hat{f}(\omega)e^{i\omega t}d\omega$$

really express the fact that f is a continuous superposition of the simple periodic waves $e^{i\omega t}$, as ω ranges over all possible frequencies. For instance, if one already knows that the dominating frequencies making up f are mostly concentrated around ω_0, say in an interval $[\omega_0 - a, \omega_0 + a]$, then this is reflected in the mathematical fact that $\hat{f}$ is mostly concentrated in the interval $[\omega_0 - a, \omega_0 + a]$. In this case, a reasonable approximation for f would be $f(t) \overset{\text{approx}}{=} \frac{1}{2\pi}\int_{\omega_0-a}^{\omega_0+a}\hat{f}(\lambda)e^{i\lambda x}d\lambda$.

If we know that *no* frequency outside the interval $[-\Omega, \Omega]$ is involved, i.e. $\hat{f}$ vanishes outside $[-\Omega, \Omega]$, then a remarkable theorem of Shannon's (the celebrated information theorist), the so-called *Shannon sampling theorem*, says that f can be completely recovered from its values at the points $t_n = \frac{n\pi}{\Omega}, n = 0, \pm 1, \pm 2, \cdots$ via the formula $f(t) = \sum_{n=-\infty}^{\infty} f\left(\frac{n\pi}{\Omega}\right) \frac{\sin(\Omega t - n\pi)}{(\Omega t - n\pi)}$. (Of course, f has to be reasonably well behaved and square integrable.) For both a heuristic as well as a rigorous justification of this fact, see pages 230–231 of [3].

Finally, we urge the reader to look up the books [2] and [3] for details regarding the theory of Fourier series and Fourier transforms and also the many important applications of this theory to other branches of mathematics, physics, electrical engineering etc.

Suggested Reading

[1] S Thangavelu. Fourier Series. *Resonance*. Vol 1, No. 10, 1996. (Also Chapter 11 in this volume)

[2] T W Körner. *Fourier Analysis*. Cambridge University Press, U.K., 1988.

[3] G B Folland. *Fourier Analysis and its Applications*. Wadsworth and Brooks/Cole, U.S.A., 1992.

A Sitaram and S Thangavelu
Indian Statistical Institute
Bangalore 560 059
Karnataka.

PART III

15

Nīlakaṇṭha Somayāji (1444–1545)

C S Yogananda

In the 19th century, the prevailing belief among the historians of science was that Mathematics and Astronomy in the Indian subcontinent had gone into hibernation after Bhāskarachārya in the 12th century. The credit for demonstrating that this was not so must surely go to Charles M Whish, Esq., a civil servant in the East-India Company. In 1832 he brought to the attention of the historians the magnificent achievements of the *Kerala School* which flourished from the 14th to 17th century. (Whish's paper, *On the Hindu Quadrature of the circle*, has been reproduced in *A Modern Introduction to Ancient Indian Mathematics*, T S Bhanumurthy, Wiley Eastern, as an appendix.) Among the major figures of this school are Mādhava (1350–1410) of Sangamagrāma, Parameśvara (1360–1455), Nīlakaṇṭha Somayāji (1444–1545) and Jyesthadēva (1500–1600), whose significant contributions to mathematics include infinite series expansions of trigonometric functions and very accurate approximations to π.

The most comprehensive work of the Kerala school available to us is the *Tantrasangraha* of Nīlakaṇṭha Somayāji along with commentaries on it by some of his followers. Fortunately, the biographical details of Nīlakaṇṭha are well recorded. He was born on Wednesday, June 14, 1444, and was a resident of Tṛkkaṇṭiyūr (Sanskritised into Śri-Kuṇḍapura), near Tirur, Ponnani taluk, South Malabar. His teachers were Ravi with whom he studied Vedānta, and Dāmodara, son of Parameśvara, who initiated him into Astronomy and the underlying mathematical principles. That Nīlakaṇṭha lived upto a ripe old age, even to become a centenarian, is attested by a contemporary reference made to him in a Malayalam work on astrology *Praśnasāra*, composed in 1542–43.

Nīlakaṇṭha's writings substantiate his knowledge of several branches of Indian philosophy and culture. It is said that he could refer to a Mīmāṁsā authority to establish his view-point in a debate and with equal felicity apply a grammatical dictum to the same purpose. Sundararāja, a contemporary Tamil astronomer, refers to Nīlakaṇṭha as *sad-darśhanī-pārangata*, one who had mastered the six systems of Indian philosophy. Another major work of Nīlakaṇṭha is his *Bhāṣya* on *Āryabhaṭīyam* of Āryabhaṭa (AD 476). The lucid manner in which difficult concepts and cryptic astronomical calculations from *Āryabhaṭīyam* are explained, the wealth of quotations, and the results of personal investigation amply justify Nīlakaṇṭha referring to his work as a *Mahābhāṣya*.

The *Tantrasangraha* of Nīlakaṇṭha Somayāji along with commentaries on it by some of his followers has been critically editied by K V Sharma. Another source book, also by K V Sharma, is *A History of the Kerala School of Hindu Astronomy;* both these books have been published by the V V B Institute of Sanskrit and Indological Studies, Punjab University, Hoshiarpur, Punjab.

C S YOGANANDA
Department of Mathematics
Indian Institute of Science
Bangalore 560 012
Karnataka.

16

Gottfried Wilhelm Leibniz, the Universalist (1646–1716)

V Chandrasekhar

The year 1996 marked the 350th birth anniversary of Leibniz (1646–1716), who towers high in the pantheon of original thinkers in the history of mankind. In a life dedicated completely to intellectual pursuits, he made fundamental and lasting contributions to an incredibly wide range of disciplines covering mathematics, logic, art of computation, mechanics, geology, jurisprudence, linguistics, philosophy and theology. This breathtaking versatility of his output puts him on the same footing as the chosen few among the scientists and philosophers (the names Archimedes, Newton, Leonardo da Vinci, Galileo immediately spring to one's mind) to whom the coveted appellation 'universalist' applies.

Gottfried Wilhelm Leibniz was born on July 1st, 1646 in Leipzig, where his father was a Professor of Moral Philosophy. During his school days he made good use of his father's library. He was enrolled in the University of Leipzig as a law student, where he was influenced by men who had revolutionized science and philosophy–Galileo, Francis Bacon, the English man of letters, Thomas Hobbes, the English philosopher, and Rene Descartes, the French mathematician and philosopher. His attempts to engage Hobbes to correspond with him to settle philosophical issues met with failure, as the latter didn't take kindly to Leibniz's interpretation of his views ("certain men are wrong in ascribing license and impiety to your hypotheses"), which was nothing more than a left-handed compliment to the Englishman.

His baccalaureate thesis, *On the Principle of the Individual* (1663) already contains the beginnings of the notion of *monad*, which was to be the fundamental unit in the philosophical system he developed later. In 1666, he anticipated the modern computer by expounding the path-breaking idea that the process of thinking could be reduced to calculation. He envisioned a kind of generalized mathematics by means of which, he claimed, "we should be able to reason in metaphysics and morals in much the same way as in geometry and analysis. If controversies were to arise, there would be no more need of disputation between two philosophers than between two accountants. For, it would suffice to take their pencils in their hands, to sit down to their slates, and to say to each other (with a friend as witness, if they liked): Let us calculate!".

The University of Leipzig refused to award him a Doctorate (because he was underaged!), which he secured later from the University of Altdorf in 1666. But he turned down the offer of a Professorship there, declaring that he had "very different things in view". He joined the court of the Archbishop of Mainz, and was involved in settling legal and political disputes.

In 1672 he was sent as an emissary to Paris to the court of Louis XIV. During his stay there for four years and in his travels to England he expanded his circle of scientific acquaintances. They included renowned men like Christian Huygens, a Dutch mathematician and astronomer who first helped with mathematics, Robert Boyle, John Pell and Tchirnhaus.

In 1673, he constructed an improved version of Pascal's arithmetic machine and presented it to the Royal Society, London. Leibniz's 'Stepped Reckoner' could perform the extraction of square roots in addition to the standard four basic operations among numbers. It must be mentioned that a completely reliable commercial model of this machine was released only in 1820. However, Leibniz had already made the crucial point that the binary system (base 2) is more natural to use than the decimal.

It was in 1673 that Leibniz made the greatest discovery of his life–calculus. The invention of calculus marked one of the most profound achievements of mankind in the History of Sciences and would have a permanent and far-reaching impact not only in mathematical sciences like geometry, physics, astronomy etc, but also in seemingly unrelated subjects like biology and economics. This work was published only in 1684 in his memoir *New Methods for the Greatest and the*

Least. Earlier in 1665, Sir Isaac Newton had also discovered calculus. He had initially contented himself by showing his work to Gregory and Collins; his work also appeared much later. This delay in the publication in both cases resulted in an unsavoury controversy between these two eminent mathematicians as to priority in the invention of calculus. The issue was settled (for the time being) on 24th April 1712 when an examining committee of the Royal Society (of which Sir Isaac was the President) pronounced a verdict in favour of Newton. But this contentious issue of priorities was never clearly resolved and was one of the most (in)famous and vexatious disputes in the 18th century. The debate still continues!

In 1676, he proposed a new formulation of mechanics called dynamics, which replaced the conservation of momentum by kinetic energy. It was around this time that he enunciated the principle that light follows the path of least resistance.

In the same year Leibniz was employed as Librarian/Councillor to the Duke of Hanover. An interesting point to note is that he possessed neither an inheritance nor a pension; thus, among his contemporaries, he was the only philosopher who had to fend for himself! In fact, it was his straitened means which motivated him to devise his calculating machine. Due to this situation, he was forced to become a jack-of-all-trades to royalty.

With his enormous mental resources, Leibniz had no difficulty in finding time for his multifarious activities. Along with his diplomatic assignments, he was simultaneously engaged in theological discussions about the reunion of the Church (a recurring theme in his life), and in carrying out his work in geology. He is acknowledged as one of the creators of this field. His point of view that the earth was at first molten found support from Buffen, a great geologist. He had a stint as a mining engineer between 1680 and 1685. During this period he worked on hydraulic presses, windmills, lamps, submarines, clocks, and a host of mechanical devices.

Between 1680 and 1683, the expansionist policies of Louis XIV, the King of France, posed an imminent threat to Germany. This provoked Leibniz to write a political pamphlet denouncing him as "The Most Christian War God". In the service of his country he devised a means of increasing the production of linen and suggested a process for the desalinization of water.

Around the same time he went back to his cherished goal of transforming human thought to symbolic manipulation. According to Bertrand Russell, Leibniz's work on mathematical logic would have been of enormous importance had he published it; and Leibniz, instead of George Boole (of Boolean Algebra fame) would be credited with being the founder of this discipline. Thus he was ahead of his times by a century and a half!

In 1685 he obtained the post of historian of the House of Brunswick and devoted himself to tracing the genealogy of the family. In this connection he travelled in Austria and Italy, where he was well received by royalty and the scientific community. His work in genealogy spurred him on this line and he set upon himself to give an account of a history of the earth, based on geological events and description of fossils. He wanted to discover facts relating to the origins and migrations of people through a study of monuments and linguistics.

In 1689 he published his essays on the dynamics of heavenly bodies and the duration of things. Leibniz was made the Librarian of Wolfenbuttel. In 1700, he founded the German Academy of Sciences in Berlin, which received royal patronage from Frederick II the Great, from 1740.

The same year he was honoured by being made a foreign member of the Academy of Sciences of Paris. In 1710, he published his famous philosophical treatise *Theodicee*, which elaborated his views on divine justice. He concludes that God chose this as the best of all possible worlds. (A sentiment very familiar to Wodehouse characters who would strongly dispute this claim!)

Peter the Great, the Tsar of Russia invited Leibniz in 1711 to Vienna. During his stay till 1714, he became a 'Reichfrat' (adviser to the empire) and was honoured with the title 'Freiherr' (baron). In 1714, he wrote his famous *Monadology* in which he synthesized his earlier work *Theodicee*.

His last years were spent in Hanover. During this time he corresponded at length with Samuel Clarke, a close associate of Newton, about the nature of space and time. In his declining years he suffered from gout and was bedridden.

This great savant passed away on November 14, 1716, leaving behind him a prodigious output (his complete works have not yet all been published). One is awed by the sheer volume of the correspondence of this supreme intellect whose mailing list contained more than 600 names!

V CHANDRASEKHAR
c/o Mr Sripathy
Chennai Mathematical Institute
92, East-coast Chambers
T Nagar, Chennai 600 017
Tamil Nadu.

17

Leonhard Euler (1707–1783)

C S Yogananda

Leonhard Euler (pronounced 'oiler') was born in Basel, Switzerland in 1707. His father was a clergyman and he wished that Leonhard would follow his footsteps though he himself had studied mathematics under Jacques Bernoulli and instructed his son in the elements of the subject. The young Euler studied under Jean Bernoulli and was associated with his sons, Daniel and Nicholas. In addition to mathematics he was broadly trained in theology, medicine, astronomy, physics, and oriental languages.

He followed Daniel Bernoulli to the St. Petersburg Academy in 1727 and was initially in the department of medicine! When Daniel Bernoulli left Russia for Basel in 1733 Euler became the Academy's chief mathematician. He married and settled down to pursue in earnest mathematical research and rear a family which ultimately included thirteen children. In 1735 he lost the sight of his right eye, due to overwork, but his misfortune in no way diminished the rate of output of his research. In 1741 Euler joined the Berlin Academy accepting an invitation from Frederick the Great and spent twenty-five years there. Unsophisticated as he was, Euler was not entirely happy at Berlin and the monarch used to refer to him as a "mathematical cyclops". He returned to the St. Petersburg Academy in 1766 and stayed there till the end. He was losing sight in his left eye by cataract and an operation was performed in 1771. Though his sight was restored for a few days, the success was short-lived and he was blind for the last seventeen years of his life. Even this tragedy did not stem the flood of research and publications which continued unabated until 1783 when at the age of seventy-six he suddenly died while sipping tea and enjoying the company of one of his grandchildren.

Euler was by far the most prolific contributor to mathematical literature and averaged about 800 pages a year; he published more than 500 books and papers during his lifetime. He worked in practically every branch of mathematics, pure and applied, and at all levels, from the most elementary to the most advanced. He was also the most successful notation builder of all times. He introduced the symbol e (e for exponential) for the base of the natural logarithm and i for $\sqrt{-1}$ and was largely responsible for popularising the symbol π to denote the ratio of the circumference of a circle to its diameter; these three symbols appear together in the equation $e^{i\pi} + 1 = 0$, a general form of which, $\cos\theta + i\sin\theta = e^{i\theta}$, is due to Euler. The notation $f(x)$ for a function of x is also due to him. One finds ubiquitous use of Eulerian symbols and terminology in geometry, algebra, trigonometry, and analysis which is why his name appears frequently in school through college-level mathematics (as well as in *Resonance*!).

Euler won the coveted biennial prize awarded by the Parisian Académie des Sciences twelve times; the essays he submitted to the académie for the contests covered a wide range of topics.

C S Yogananda
Department of Mathematics
Indian Institute of Science
Bangalore 560 012
Karnataka.

18

Jean Baptiste Joseph Fourier (1768–1830)

A Sitaram

Administrator, Egyptologist, engineer, mathematician, physicist, revolutionary, soldier, and teacher; incredible as it may sound, Jean Baptiste Joseph Fourier was all these! Born on March 21, 1768 in Auxerre, France, orphaned at the age of ten, Fourier had a brilliant school career in the local Benedectine school in his home town. His life was rich and varied: member and president of the local revolutionary committee during his youth, student at the Ecole Normale, teacher at the Ecole Polytechnique (where he succeeded the great Lagrange in the Chair of Analysis and Mechanics), imprisoned several times thanks to rapidly shifting political ideologies in Paris, member of the Napoleonic expedition to Egypt, important administrator in Egypt as well as in France under both Napoleon and the King of France (he held, brilliantly, the post of Prefect of the Department of Isere centred around Grenoble for fourteen long years), and finally, as recognition of his distinguished service to science, he was elected permanent mathematical secretary of the French Academy of Sciences in 1822. His scientific and mathematical achievements are, of course, legion. He derived the partial differential equation that governs heat conduction, and, to study the problem of heat conduction, systematically developed the subject which later came to be known as Fourier analysis – a subject which was in its infancy during his time. Though there were many other important players, Fourier is justly considered the father of Fourier analysis. Some of his great discoveries are contained in his celebrated treatise *'Théorie analytiqúe de la chaleur'* published in 1822. Fourier also made important contributions to probability theory, statistics, and mechanics. What is perhaps not so well known is that some of the ideas of the modern theories of optimization and linear programming date back to Fourier in the years 1823 and 1824. (See the article by V Chandru and M R Rao in *Resonance*, October 1998). One of the greatest mathematical analysts of his time, this is what Fourier had to say about his own discipline (see pages 558–559 of *'Fourier Analysis'* by T W Körner, CUP 1988): " ... mathematical analysis is as extensive as nature herself. ... This difficult science grows slowly, but once ground has been gained it is never relinquished. It grows, without ceasing, in size and strength, amid the constant error and confusion of the human spirit. ... it follows the same path when applied to all phenomena and interprets them all in the same language as if to attest to the unity and simplicity of the design of the universe and to make still more evident that unchanging order which presides over all natural laws."

A Sitaram
Indian Statistical Institute
R V College Post
Bangalore 560 059
Karnataka.

19

Carl Friedrich Gauss (1777–1855)

D Surya Ramana

Carl Friedrich Gauss, whose mathematical work in the 19th century earned him the title of Prince among Mathematicians,[1] was born into the family of Gebhard Dietrich Gauss, a bricklayer at the German town of Brunswick, on 30 April 1777.

Gauss was a child prodigy whose calculating prowess showed itself fairly early in his life. Stories of his childhood are full of anecdotes involving his abilities that first amazed his mother, then his teachers at school, and much more recently his biographers to the present day. For instance, when he was barely ten years old he surprised his teachers by calculating the sum of the first hundred natural numbers using the fact that they formed an arithmetical progression.

While his talents impressed his mother Dorothea, they caused a great deal of anxiety to his father. Apparently afraid of losing a hand that would otherwise ply some useful trade, he very reluctantly permitted his son to enroll in a gymnasium (high school) at Brunswick. Gauss studied here from 1788–1791.

Perhaps conscious of the fact that his father would not sponsor his studies anymore, Gauss began to look for a patron who would finance his education beyond the gymnasium. Encouraged by his teachers and his mother, Gauss appeared in the court of Duke Carl Wilhelm Ferdinand of Brunswick to demonstrate his computational abilities. The Duke immediately recognised the genius of his subject

[1] So varied was the mathematical work of Gauss that this article is limited in scope by the author's own special interests.

and granted the fourteen-year-old Gauss 'whatever means should be needed for the continued training of such a gifted person'.

When he was barley ten years old, Gauss surprised his teachers by calculating the sum of the first hundred natural numbers using the fact that they formed an arithmetical progression.

In 1792 Gauss entered the Collegium Carolinum and also made the first important studies of his mathematical life. These studies, which were on the iterative calculation of arithmetic and geometric means starting with two positive numbers, had already led him to the study of elliptic functions, a subject that would develop much later at the hands of Jacobi and Abel. In the following year, when he was merely fifteen, he had empirically conjectured an asymptotic formula for the number of primes $\pi(n)$ less than a given integer n. This formula, which today is known as *the prime number theorem*, was proved only a century later by de la Vallée Poussin and Hadamard. However, the most significant contribution of his days at the Collegium Carolinum was the construction of the regular 17-sided polygon using only a ruler and compass. This simply-stated problem in elementary geometry, which had remained unsolved since the time of the Greeks, had beneath its surface certain algebraic and number theoretic notions that needed the genius of Gauss to come to light. Indeed, not only had he constructed the regular 17-sided polygon, but had also determined precisely which regular polygons could be constructed by ruler and compass alone.

It is said that upto the point when he solved this problem, Gauss had not decided whether he would pursue mathematics or humanities, which was economically more beneficial during his time. His solution to this two thousand year-old Greek problem elated him so much that he decided to pursue mathematics.

The fact that he had discovered the principle based on which one "divides a circle into 17 parts and so forth" is also the first entry in his mathematical diary. The entry is dated 30 March 1796. This diary, which he kept throughout his life, contained many of those ideas that had come to him but which he had not been able to work out fully and publish. His notings in this diary, which was discovered among his papers after his death, have been the basis for several investigations even upto recent times.

Simultaneously with noting this discovery in his diary he also announced it in a literary journal. In this announcement he says that the discovery of the ruler and compass construction for the 17-sided polygon "is really only a corollary to a greater theory which will be published soon". This greater theory appeared in his book *Disquitiones Arithmeticae or Arithmetical Investigations*, which he published in mid-1801 at the age of 25.

The subject matter of this book is number theory, the branch of mathematics that deals with the properties and relations between integers. In the *Disquitiones*, Gauss developed all the number theory of his time from first principles, and in fact went far ahead.

Gauss' Construction of the 17-gon

To construct a regular 17-gon geometrically (using ruler and compass) we must find the seventeenth roots of unity and then locate the roots on the unit circle geometrically. The seventeenth roots of unity are found by solving the equation:

$$x^{17} - 1 = 0.$$

The roots are 1, $\cos(2r\pi/17) \pm i \sin(2r\pi/17), r = 1, 2, \ldots, 8$.
Let $y_r = \cos(2r\pi/17), r = 1, 2, \ldots, 8$. Then using the fact that the sum of the roots of the equation is zero, we have

$$2(y_1 + y_2 + \ldots + y_8) + 1 = 0, \text{ i.e., } y_1 + y_2 + \ldots + y_8 = -1/2.$$

From the standard conversion formulae we get $2y_i y_j = y_{i+j} + y_{|i-j|}$. Further, pairs of $y_1, y_2, \ldots, y_8$ have a remarkable cyclic property which is the key to the construction:

$$2y_1y_4 = y_3 + y_5, 2y_3y_5 = y_2 + y_8, 2y_2y_8 = y_6 + y_7, 2y_6y_7 = y_1 + y_4$$

Let $y_1 + y_4 = 2\alpha, y_3 + y_5 = 2\beta, y_2 + y_8 = 2\gamma, y_6 + y_7 = 2\delta$. Then,

$$\alpha + \beta + \gamma + \delta = -1/4, \text{and } \alpha\gamma = (y_1 + y_4)(y_2 + y_8)/4$$
$$= (y_1 + y_2 + \ldots + y_8)/8 = -1/16.$$

Similarly, $\beta\delta = -1/16$. Also

$$4\alpha\beta = (y_1 + y_4)(y_3 + y_5) = (2y_1 + y_2 + 2y_4 + y_6 + y_7 + y_8)/2 = -\frac{1}{4} + \alpha - \beta$$

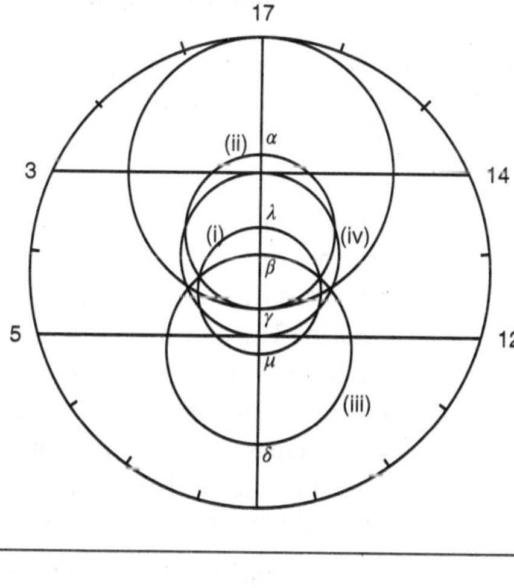

Similarly, $16\beta\gamma = -1 + 4\beta - 4\gamma$. So also for $16\gamma\delta$ and $16\delta\alpha$.
Let $\alpha + \gamma = 2\lambda, \beta + \delta = 2\mu$. Then

$$\lambda + \mu = -1/8, 4\lambda\mu = \alpha\beta + \beta\gamma + \gamma\delta + \delta\alpha = -1/4.$$

Thus, λ, μ satisfy the quadratic equation: $y^2 + (1/8)y - (1/16) = 0$. The pairs $\{\beta, \delta\}, \{\alpha, \gamma\}$ satisfy quadratic equations involving λ and μ, and the pairs $\{y_6, y_7\}, \{y_2, y_8\}, \{y_3, y_5\}$, and $\{y_1, y_4\}$ satisfy quadratic equations involving α, β, γ, and δ.

The roots of a quadratic equation of the type $y^2 - 2ay + b = 0$ are the ordinates of the points, where the $x = 0$ cuts the circle $x^2 + y^2 - 2ay + b = 0$. Thus we can locate the points $y_r, r = 1, 2, \ldots, 8$, by drawing the corresponding circles.
These steps can be summarised in the following algorithm.

1) Draw a unit circle.
2) Draw a circle centre $(0, -1/16)$ passing through the points $(\pm 1/4, 0)$ cutting the y-axis at λ and μ. (Circle (i))
3) Draw two circles centres $(0, \lambda)$ and $(0, \mu)$ passing through the points ($\pm$ 1/4,0) cutting the y-axis at α, γ and β, δ respectively. (Circles (ii) and (iii))
4) Draw the circle on $(0, 1)$, $(0, \gamma)$ as diameter, cutting the x-axis at $(\sqrt{-\gamma}, 0)$. (Circle (iv))
5) Draw the circle with $(0, \beta)$ as centre, passing through $(\sqrt{-\gamma}, 0)$.
6) Draw the lines parallel to the x-axis touching this circle at y_3, y_5, y_{12}, y_{14}.
7) Bisect the intercepted arcs on the unit circle and mark off the vertices.

RENUKA RAVINDRAN, Department of Mathematics
Indian Institute of Science, Bangalore 560 012, India.

Gauss proved that a regular polygon with n sides, where n is an odd integer, can be constructed using a ruler and compass alone if and only if n can be expressed as a product of distinct Fermat primes.

The high point of the initial part of the book, comprising the first four chapters, is the first complete proof of the so-called *law of quadratic reciprocity*. This law had been conjectured variously by Euler, Lagrange and Legendre, though none of them was able to prove it. Unaware of all this, Gauss had independently conjectured the law, and shortly before his nineteenth birthday, obtained the first correct proof. Gauss seemed more enamoured of the law than of the proof he obtained. He called it 'the fundamental law' and 'the golden law' and spent many years generalising it. His attempts at this generalisation, which were followed up by his disciple Eisenstein and thereafter by other great number theorists of the nineteenth and twentieth centuries, resulted in the creation of an important branch of present day number theory called *Class Field Theory*.

The second part of his book, comprising the fifth and sixth chapters, dealt with his theory of quadratic forms and its applications. A quadratic form, in this context, is an expression of the type $ax^2 + bxy + cy^2$, where a, b, c are given integers. The principal question at the time of Gauss' investigation was to determine, given an integer m and a quadratic form as above, if there exist integer values x, y such that $ax^2 + bxy + cy^2 = m$.

Like with quadratic reciprocity, Gauss had come to this question on his own. In his study on quadratic forms, which is often considered the best of his published works, he went far ahead of his time. It is this work that led him to suspect the existence of generalisations of the law of quadratic reciprocity.

The seventh and final chapter of his book was devoted to the development of the theory that Gauss had announced in connection with the construction of the 17-sided regular polygon. Inaugurating the study of what are now called cyclotomic numbers, Gauss proved that a regular polygon with n sides, where n is an odd integer, can be constructed using a ruler and compass alone if and only if n can be expressed as a product of distinct Fermat primes.[2]

By the time this book was published, Gauss had also been awarded a doctorate in mathematics. For his doctorate, which was awarded in absentia by the University of Helmstedt in July 1796, he had submitted a thesis giving a new proof of what is called *the Fundamental Theorem of Algebra*.

However, the event that really brought him recognition was his solution to a problem that was, quite literally, of astronomical proportions.

On Jan 1, 1801, about six months before the publication of *Disquitiones*, the Italian astronomer Joseph Piazzi had sighted the asteroid Ceres. For about forty days he carried out observations on it. In this period it swept an angle of about nine degrees and thereafter disappeared into the light of the Sun. While the philosopher Hegel ridiculed the astronomers' quest for more planets and asteroids, (he thought he had logically proved that there could be only seven of them), Gauss calmly went about the problem of computing the entire orbit of Ceres with the minimal amount of data that Piazzi had gleaned. In very little time he had computed the entire orbit and predicted exactly

In very little time he had computed the entire orbit and predicted exactly the position at which the asteroid would be seen again. Ceres appeared in the skies exactly as Gauss had said it would and he was immediately recognised a master.

[2] A Fermat prime is a prime integer that can be expressed as $2^{2^k} + 1$, for some integer k.

the position at which the asteroid would be seen again. Ceres appeared in the skies exactly as Gauss had said it would, and he was immediately recognised a master.

Gauss got married to Johanna Osthoff, the daughter of a tannery owner at Brunswick in 1805. He was still without a permanent job. Despite his fame, his financial position still depended on the stipend from the Duke of Brunswick. To make matters worse, the Duke was engaged in a battle with the then formidable Napoleon Bonaparte whose eventually futile attempt at conquering the world had brought him upto Brunswick. It appeared that the Duke was destined to live longer in the memory of mathematicians rather than on the battlefield. For in October 1806 he was grievously injured and died soon after. His death affected Gauss deeply. However, shortly after this tragedy, the University of Göttingen offered him a professorship of Astronomy which he accepted in November 1806. He remained in that position for the rest of his life.

Matrimony was quick to bestow its favours on Gauss. In 1806 a son was born to him and two years later a daughter. But then tragedy struck again. In 1809, after a third delivery, his wife passed away and in a few months their infant son also died, leaving Gauss very depressed. In the following year Gauss married Minna Waldeck, the closest friend of his late wife in her last years and by 1816 he was a happily settled man with an awesome reputation, a permanent position and a family with five children, three from his second marriage.

Until 1816 owing to his studies in astronomy Gauss investigated the estimation of errors in observation. In the process, he developed the method of least squares, the law of normal distribution (which is also called Gaussian distribution) and also perturbation methods for studying planetary orbits.

The Gauss–Bonnet theorem states that the total curvature of a surface without boundary is a topological invariant.

Starting in 1820 he turned to differential geometry of surfaces motivated by his work on geodesy. In 1827 he published his results in a paper that has become a landmark in the subject. In this paper he introduced the notion of what is now called the Gaussian curvature of a surface. In an earlier work on this subject, Euler had introduced two curvatures for a surface embedded in space. However, the Eulerian curvatures depended on the way in which the surface was embedded in space. On the other hand, Gaussian curvature does not change as long as one deforms the surface without stretching or tearing. This is a consequence of Gauss' *Theorema Egregium* (outstanding theorem) that was proved in this paper. He proved that the sum of the angles of a triangle on a surface differs from 180° by an amount proportional to the total curvature of the triangle. This later led to the Gauss–Bonnet theorem which states that the total curvature of a surface without boundary is a topological invariant. Indeed some feel that this paper shows that Gauss had a clear notion of non-Euclidean geometry far ahead of Bolyai and Lobachevsky. A more practical consequence of Gauss' work was a proof of the impossibility of making constant scale maps of the Earth.

In Brunswick, Germany, in 1780, a stonemason was calculating the wages due to his workmen at the end of the week. Watching him was his three-year-old son. "Father," said the child, "the reckoning is wrong." The boy gave a different total which, to everyone's surprise, was correct. No one had taught the lad any arithmetic. The father had hoped his son would become a bricklayer, but thanks to his mother's encouragement, the boy, Carl Friedrich Gauss, became one of the greatest mathematicians in history.

Indeed some feel that Gauss had a clear notion of non-Euclidean geometry far ahead of Bolyai and Lobachevsky.

By the year 1831 his varied interests had taken him to physics. Upon his recommendation, the experimental physicist Wilhelm Weber was appointed a professor at Göttingen. Gauss and Weber formed an unbeatable combination. By the year 1833 they had invented the electromagnetic telegraph and built a theory for explaining terrestrial magnetism. Indeed they had also investigated the question of determining the strength of the earth's magnetic field. It is in honour of these studies that the SI unit for magnetic flux density is called a *weber* and one ten thousandth of 1 weber, which roughly corresponds to the terrestrial magnetic flux, is called a *gauss*. During these years (1833–1855), Gauss also worked on potential theory and optics.

Gauss passed away on 23 February, 1855, at the age of 78—a rather long life for a genius of his calibre and times. Whether the Gods loved him or not, he had surely earned the respect of mathematicians and scientists of his times and those that followed, including, by now, perhaps the reader.

D Surya Ramana
Institute of Mathematical Sciences
CIT Campus, Taramani
Chennai 600 113
Tamil Nadu.

20

Sonya Kovalevskaya (1850–1891)

The First Woman Mathematician of Modern Times

Mythily Ramaswamy

Sonya Kovalevskaya was born as Sofya [1] Vasilievna Krukovskaya on 15th January, 1850 in Moscow in an aristocratic family. By nature she was an intense and serious person. An early indication of her talents was her ability to make sense of the sine function all by herself. It was then, that her conservative father allowed her to take some calculus lessons from Prof. Strannoliubsky in St. Petersburg.

[1] Sometime in her adult life, Sofya acquired the name Sonya by which she is better known.

Russian society being very conservative at that time, women had no hopes of higher education or a career. Many young women tried to go to the west but could travel only with the permission of their husbands or fathers. In some extreme cases, a young woman entered into a "marriage" for namesake with a liberal "husband" and the couple chaperoned a group of her friends to the west. Vladimir Kovalevsky, a liberal and publisher of many scientific texts "married" Sofya in 1868 and the couple escorted her sister Anyuta to the west, in early 1869.

Sofya and Vladimir settled down in Heidelberg where she studied for a year with du Bois-Reymond and Konigsberger, both students of Karl Weierstrass, one of the giants of mathematics. This too was possible only after her strenuous efforts at getting the administration to agree that she could attend courses with the permission of the professors. In the fall of 1870, she left for Berlin to seek out her teacher's teacher, Weierstrass. As he was forbidden by university rules to allow her to attend his lectures, he started giving her private lessons. Thus Sofya was able to catch up on her mathematical education, in spite of turbulent events in her personal life. In a period of 18 months, she wrote 3 dissertations under Weierstrass's direction. He appealed to the somewhat more liberal University of Göttingen for her doctoral degree which she received in 1874 to become the first woman Ph.D.

After their return to Russia in 1875, Sofya and Vladimir (who had a Ph.D. in Geology) were unable to find any suitable position and hence turned to real estate business investments which eventually failed. Meanwhile in 1878, her daughter, Fufa was born.

In March 1881, Tsar Alexander II was assassinated and the couple, with their radical ties, found it safer to move to the west. Sofya settled down in Berlin to work on Lame's equations. Vladimir returned to Russia and was getting deeper into financial trouble and finally committed suicide in April 1883. This was a rude shock to Sofya. In spite of that, she managed to finish her work. Meanwhile Mittag-Leffler had arranged a post for her in Stockholm University. She felt that it was her moral obligation to clear Vladimir of the charges against him. So she returned to Russia and managed to establish his innocence. Then leaving Fufa with her friend Julia she sailed to Stockholm in November 1883, to become the first woman professor of mathematics. The following years were quite productive mathematically. Her work on Euler's equations fetched her the Bordin Prize from the Paris Academy in 1888. But in February 1891, her life came to a sudden end after a brief illness.

In her thesis, she had proved an important theorem on the existence and uniqueness of an analytic solution for partial differential equations with analytic coefficients (now known as the Cauchy–Kovalevskaya theorem). Her thesis also contained results on degenerate Abelian integrals which reduce to elliptic integrals, and on the shape of Saturn's rings. The important work which fetched her the Bordin prize was on Euler's equations describing the motion of a heavy rigid body, with one point fixed like a top or a pendulum. Her other work includes the one on Lame's equations for light refraction in crystals.

In the field of literature, in the last five years of her life she produced a memoir of her childhood and adolescence, a novel called *A Nihilist Girl,* two plays, a small body of verse, a collection of short stories and some essays.

Thus, in spite of the uncertain political atmosphere and a turbulent personal life, Sofya Kovalevskaya managed to pursue a successful mathematical career thanks to her talents and determination. Of course, the influence of Weierstrass and the extraordinary efforts of Mittag–Leffler are equally important factors that helped her. She is recognized as the first woman mathematician of modern times, who, through her achievements, opened many doors for other women in the field of mathematics.

Suggested Reading

E T Bell. *Men of Mathematics.* Simon and Schuster, 1937.

B Stillman. *A Russian Childhood*. Springer Verlag, 1978 (translated, edited and introduced by Stillman from Sofya's original, *Memories of Childhood in Russia).*

A H Koblitz. *A Convergence of Lives, Sofya Kovalevskaya: Scientist, Writer, Revolutionary*. Birkhäuser, 1983.

Roger Cooke. *The Mathematics of Sonya Kovalevskaya*. Springer Verlag, 1984.

MYTHILY RAMASWAMY
TIFR Centre, IISc Campus
Bangalore 560 012
Karnataka

"....*You are surprised at my working simultaneously in literature and in mathematics.One of the foremost mathematicians of our century says very justly that it is impossible to be a mathematician without also being a poet in spirit.It seems to me that the poet must see what others do not see, must see more deeply than other people.And a mathematician must do the same....* "

— *Sonya in a letter to a friend in 1890*

21

The Man Who Knew Infinity

Untutored Genius

R Tandon

It is not very often that one sees the biography of a pure mathematician—not a mathematician like Newton who was also a physicist, but a pure mathematician who worked in number theory, probably the purest realm of mathematics. Ramanujan's work may today have some applications in particle physics or in the calculation of π upto a very large number of decimal places but it would be misleading to stress this aspect of his work. Ramanujan did mathematics for its own sake, for the thrill that he got in seeing and discovering unusual relationships between various mathematical objects. The oft repeated anecdote about Ramanujan finding the number 1729 interesting because it is the smallest number that can be written as a sum of two cubes in two different ways, $10^3 + 9^3, 12^3 + 1^3$ bears repetition only in that it gives a flavour of the kind of mathematics that Ramanujan was interested in—the kind of mathematics about which the sceptic will say, "so what!" and surely Ramanujan would have said, "so what"!

Ramanujan was born in a poor Tamil Brahmin family that resided in the town of Kumbakonam. He attended school there and did averagely well. While in school he came across a book entitled *A Synopsis of Elementary Results in Pure and Applied Mathematics* by George Carr. This book is just a compendium of results on integrals, infinite series and other mathematical entities found in analysis. Yet it left a lasting impression on Ramanujan; in fact it virtually determined his mathematical style. He would later write mathematics as a string of results without proof or with the barest outline of a proof.

After school Ramanujan was hooked on mathematics. He spent all his time with his head bent over a slate, working on problems in number theory that interested him, and neglected everything else. The result was that he could never get through another examination. An early marriage, as was usual in those times, led to a frantic search for a job to earn an income. He became a clerk in the Madras Port Trust with the help of some well-wishers. In the meantime Ramanujan kept showing his results to various people who he thought would be interested or would help him get a job that would give him a lot of time to do mathematics. He wrote to a couple of well-known British mathematicians giving a list of some of the results he had obtained. They ignored him—thought he was a crank! Finally he wrote to one of the most distinguished English mathematicians of the time—a person who had done a lot of work on number theory—G H Hardy. Hardy arranged for Ramanujan to come to Trinity College, Cambridge where he and Ramanujan met almost daily discussing mathematics for about three years. Starting from the early life of Ramanujan, Kanigel describes all this and much more in his beautiful story on Ramanujan.

Kanigel's book is not only a biography of Ramanujan, but contains a minibiography of Hardy as well. On Hardy, Kanigel is brilliant. Though basically a conservative Englishman, Hardy was a revolutionary in the world of British mathematics. (The ethos of British mathematics of the time was determined by the famous Tripos examination. It is not unusual even today to hear someone of our father's generation boasting that so and so was a Wrangler—that is, had acquired a high rank in the Tripos examination. To Hardy however the Tripos was an anachronism—it had outlived its usefulness.) Kanigel writes about all this with a great deal of perception, which is especially admirable in view of the constraints he worked with. To quote Kanigel himself: "In writing the life of Ramanujan, I faced the barriers of two foreign cultures, a challenging discipline, and a distant time. As I am expert in none of these, I owe a debt of gratitude to the many persons who have helped me surmount those barriers—who have consented to

---

Review of the book *The Man Who Knew Infinity–A Life of the Genius Ramanujan* authored by Robert Kanigel.

interviewing, spent hours explaining recondite areas of mathematics or Indian cultural life, guided me to out of the way documents in libraries and archives, read and criticized early drafts, befriended me in England and India—and, back in Baltimore, offered a supportive hand or word of advice".

Kanigel covers in a most balanced manner two controversies that are usually associated with Ramanujan—one: whether Ramanujan drew upon divine or some sort of mystical inspiration to come up with his ideas, or whether, as was maintained by Hardy, he was just a hard-working, exceptionally original and creative person like other great mathematicians. The second controversy concerns the medical cause of Ramanujan's death.

So, what did I get from Robert Kanigel's book on Ramanujan? I think it reinforced my belief in the universality of the language of science. Here were two totally dissimilar people both culturally and temperamentally—from two totally different backgrounds. Each knew very little about aspects of the other's personal life, and when Hardy learnt later of Ramanujan's personal problems they came as a complete surprise to him. Of course, Hardy himself was too reserved to let Ramanujan get even a whiff of his own concerns. But when they met, which they did almost every day for nearly three years, they were on exactly the same wavelength—they spoke exactly the same language—they were totally *intimate* with each other in the language of mathematics.

I would like to end by quoting from the last few pages of the book. It was 1936; Harvard University was celebrating the 300th anniversary of its founding and as part of the celebrations it was honoring sixty eminent men of letters and science. The list included Einstein, Heisenberg, Piaget and Jung. They were also honoring Hardy. Hardy gave a series of lectures that later came out as a book entitled *Ramanujan: Twelve lectures on subjects suggested by his life and work.* Hardy said at the beginning of his lectures, and I quote from Kanigel's book:

"I have to form myself, as I have never really formed before, and to try to help you to form, some sort of reasoned estimate of the most romantic figure in the recent history of mathematics, a man whose career seems full of paradoxes and contradictions, who defies almost all canons by which we are accustomed to judge one another and about whom all of us will probably agree in one judgement only, that he was in some sense a very great mathematician".

And then Hardy began to speak about his friend Ramanujan.

Kanigel tries to do the same. Hardy concentrated on the technical aspects of Ramanujan's work. Kanigel in his own meticulously researched style tells us more about Ramanujan the man.

R Tandon
Department of Mathematics and Statistics
University of Hyderabad
Hyderabad 500 046
Andhra Pradesh.

"In South India today, everyone has heard of Ramanujan. College professors and bicycle rickshaw drivers alike know his story, at least in sketchy outline, just as everyone in the West knows of Einstein. Few can say much about his work, and yet something in the story of his struggle for the chance to pursue his work on his own terms compels the imagination, leaving Ramanujan a symbol for genius, for the obstacle it faces, for the burden it bears, for the pleasure it takes in its own existence".

ROBERT KANIGEL in *"The Man Who Knew Infinity"*

22

Harish-Chandra

R Tandon

Harish-Chandra was born in Kanpur on the 11th of October 1923 to Chandrakishore and Chandrarani. The men in the family were professionals (lawyers, engineers), and on his mother's side, the women had obtained convent education even during the 19th century. Chandrakishore was an irrigation engineer who spent much of his time in the districts of U.P. Influenced by the times, he became a staunch Gandhian and dropped his surname because of his opposition to caste. Harish-Chandra initially studied with private tutors until the age of nine, when he was directly enrolled in the seventh class. During his childhood and youth, he liked to paint, although later he gave up painting altogether.

Contact with Prof. K S Krishnan at Allahabad University during his MSc changed his future. Prof. Krishnan was quick to recognise Harish-Chandra's extraordinary mathematical skills and encouraged him to take up physics as a career. Harish-Chandra was persuaded to go to the Indian Institute of Science, Bangalore where he worked with Homi Bhabha during 1943–45. The first six months of his stay in Bangalore were spent with old friends from Allahabad, Mr and Mrs G T Kale. It was here that he got to know their daughter Lalitha, and many years later when he returned to India on a visit in 1952, she became his wife. The war in Europe over, Harish-Chandra left for Cambridge to work with Paul Dirac. This was the genesis of a long friendship and Harish-Chandra credits Dirac with being one of the most important influences on his life. For his PhD thesis

Dirac suggested a study of the infinite dimensional representations of the Lorentz group. Harish-Chandra completed this work about the same time as I M Gelfand and M A Naimark in the erstwhile Soviet Union, and V Bargmann in the United States.

Harish-Chandra spent the year 1947–48 in USA as Dirac's assistant. His earlier doctoral work together with a year of attending lectures by Claude Chevalley and Emil Artin—and the recipe for the conversion of Harish-Chandra from a physicist to a mathematician was complete. He always regarded physics with a certain amount of awe and claimed that you needed to have a sixth sense to be a successful physicist, something he thought he lacked. He preferred the unfettered freedom of mathematics.

For the next 15 years Harish-Chandra was at Harvard, Princeton or Columbia. There was a break of a year in 1952–53 which he spent at the Tata Institute of Fundamental Research, briefly flirted with, but then abandoned the idea of returning to India. During these years he was building the monument for which he is famous, the general theory of semi-simple Lie groups and the construction of the Plancherel measure for such groups. He became a professor at the Institute for Advanced Study, Princeton in 1963 and he announced that he had the general Plancherel formula at the International Congress of Mathematicians, Moscow 1966. *Then* followed several long papers to support his claim. He then turned his attention to reductive groups over the *p*-adics because "there is no reason why the reals should be more important than any other completion of the rationals." In the words of R P Langlands "Harish-Chandra was one of the outstanding mathematicians of his generation, an algebraist and analyst, and one of those responsible for transforming infinite-dimensional group representation theory from a modest topic on the periphery of mathematics and physics into a major field central to contemporary mathematics." Harish-Chandra worked incredibly hard, often all night through, night after night. This took its toll and his health deteriorated. He published just one joint paper with A Borel in 1960 at the time of his first serious illness.

Harish-Chandra won the Coles Prize of the American Mathematical Society in 1954, the Srinivasa Ramanujan Medal of INSA in 1974, became a Fellow of the Royal Society in 1973 and a member of the National Academy of Sciences, USA in 1981. He was made the IBM von Neumann Professor at the Institute for Advanced Study in 1968.

During a talk on the 80th birthday celebration for Dirac, Harish-Chandra said "I have often pondered over the roles of knowledge or experience on the one hand, and imagination or intuition, on the other, in the process of discovery. I believe that there is a certain fundamental conflict between the two, and knowledge, by advocating caution, tends to inhibit the flight of imagination. Therefore a certain naiveté, unburdened by conventional wisdom, can sometimes be a positive asset. I regard Dirac's discovery of the relativistic equation of the electron as a shining example of such a case." Harish-Chandra, like most creative mathematicians, believed in the dictum "Mathematics is like music; you must practise it in order to be able to teach it."

Harish-Chandra died on 16 October 1983.

R TANDON
Department of Mathematics and Statistics
University of Hyderabad
Hyderabad 500 046
Andhra Pradesh

"It is difficult to communicate the grandeur of Harish-Chandra's achievements and I have not tried to do so.The theory he created still stands—if I may be excused a clumsy simile—like a Gothic cathedral, heavily buttressed below but, in spite of its great weight, light and soaring in its upper reaches, coming as close to heaven as mathematics can.Harish, who was of a spiritual, even religious cast, and who liked to express himself in metaphors, vivid and compelling, did see, I believe, mathematics as mediating between man and what one can only call God.Occasionally, on a stroll after a seminar, usually towards evening, he would express his feelings, his fine hands slightly upraised, his eyes intent on the distant sky; but he saw as his task not to bring men closer to God but God closer to men.For those who can understand his work and who accept that God has a mathematical side, he accomplished it."

R P LANGLANDS—*Harish-Chandra: In Memoriam,* Princeton, April 23, 1984

"In the austere simplicity and uncompromising nature of his approach to life, in his preference for solitary and profound reflection, and in his awesome capacity to discern and persevere after distant goals, he resembled the legendary figures from his country's ancient past.And like them, he came to be quite detached about his achievements as well as his failures.This detachment was not a false modesty; like many great men Harish-Chandra was fully conscious of his gifts and what he could do with them.It was rather a deeper humility, whose origin lay in a conviction that science was a collective endeavour and that any life is but a fragment in a larger fabric."

V S VARADARAJAN—*Harish-Chandra: In Memorium*, Princeton, April 23, 1984

When and Where the Articles Appeared in Resonance

PART I

1. *Mathematical Induction—An Impresario of the Infinite*, Feb 98, 3(2), 69–76 : B Sury.
2. *Nakula's Extraordinary Talent—Mathematics from the Mahabharata*, Jan. 97, 2(1), 44–52 : C Musili.
3. *Transfinite Numbers—What is infinity?*, Mar. 97, 2(3), 58–68 : S M Srivastava.

PART II

4. *Introductory Remarks*: V Pati and A Sitaram.
5. *An Elementary Problem that Interested Ramanujan*, Dec. 96, 1(12), 60–62 : S A Shirali.
6. *The Vacillating Mathematician—The Deterministic Version*, (excerpts from the article "The Vacillating Mathematician"), Jan. 97, 2(1), 16–19, 24 : K B Athreya.
7. *A Route to Square Roots and Cube Roots*, Nov. 97, 2(11), 84–94 : S A Shirali.
8. *On the Infinitude of Primes—Euler's Proof*, Mar. 96, 1(3), 78–95 : S A Shirali.
9. *Nīlakaṇṭha, Euler and π*, May 97, 2(5), 29–43, and Nov. 97, 2(11), 112 : S A Shirali.
10. *Stirling's Formula*, (excerpts from the article "The Normal Distribution"), Jun. 97, 2(6), 18–19 : S Ramasubramanian.
11. *Fourier Series—The Mathematics of Periodic Phenomena*, Oct. 96, 1(10), 44–55 : S Thangavelu.
12. *The Isoperimetric Problem*, Sept. 97, 2(9), 65–68 : A Sitaram (reprinted from Bona Mathematica, Dec. 96).
13. *From Calculus to Wavelets: A New Mathematical Technique—Wavelet Analysis*, Apr. 97, 2(4), 25–37 : G B Folland.
14. *From Fourier Series to Fourier Transforms*, Oct. 98, 3(10), 3–5 : A Sitaram and S Thangavelu.

PART III

15. *Nīlakaṇṭha Somayāji (1444–1545)*, May 97, 2(5), 3 : C S Yogananda.
16. *Gottfried Wilhelm Leibniz, the Universalist*, Mar. 97, 2(3), 2–4 : V Chandrasekhar.
17. *Leonhard Euler (1707–1783)*, May 97, 2(5), 4 : C S Yogananda.

18. *Jean Baptiste Joseph Fourier (March 21, 1768–May 16, 1830)*, Oct. 98, 3(10), 101 : A Sitaram.
19. *Carl Friedrich Gauss*, June 97, 2(6), 60–67 : D Surya Ramana (with a boxed-item by R Ravindran on Gauss' construction of the 17-gon).
20. *Sonya Kovalevskaya (Sonya Krukovskaya)*—The First Woman Mathematician of Modern Times, Sept. 96, 1(9), 3–4 : Mythily Ramaswamy.
21. *The Man Who Knew Infinity: Untutored Genius*, (Review of the book "The Man Who Knew Infinity—A Life of the Genius Ramanujan"), Dec. 96, 1(12), 76–78 : R Tandon.
22. *Harish Chandra*, Aug. 96, 1(8), 3–4 : R Tandon.

Index